Studies in Fuzziness and Soft Computing 304

For further volumes:
http://www.springer.com/series/2941

Zongmin Ma and Li Yan (Eds.)

Advances in Probabilistic Databases for Uncertain Information Management

Springer

Editors
Zongmin Ma
College of Information Science and Engineering
Northeastern University
Shenyang
China

Li Yan
School of Software
Northeastern University
Shenyang
China

ISSN 1434-9922 ISSN 1860-0808 (electronic)
ISBN 978-3-642-37508-8 ISBN 978-3-642-37509-5 (eBook)
DOI 10.1007/978-3-642-37509-5
Springer Heidelberg New York Dordrecht London

Library of Congress Control Number: 2013935339

Printed on acid-free paper

Springer is part of Springer Science+Business Media (www.springer.com)

Preface

Databases are designed to support the data storage, processing, and retrieval activities related to data management in information systems. Database management systems provide efficient task support and tremendous gain in productivity is hereby accomplished using these technologies. In addition, being the de-facto standard for data representation and exchange over the Web, XML (Extensible Markup Language) has been widely and deeply applied in many business, service, and multimedia applications, and a large volume of data is managed today directly in XML format.

While traditional database models provide powerful data modeling and data processing capabilities, they may suffer from some inadequacy of necessary semantics. One of the major areas of database research has been the continuous effort to enrich existing database models with a more extensive collection of semantic concepts in order to satisfy the requirements in the real-world applications. In real-world applications, information is often imperfect, and human knowledge and natural language have a big deal of imprecision and vagueness. Traditional database models assume that the models are a correct reflection of the world and further assume that the stored data is known, accurate and complete. It is rarely the case in real life that all or most of these assumptions are met. One of some inadequacy of necessary semantics that traditional database models often suffer from can hereby be generalized as the inability to handle imprecise and uncertain information. For this reason, imprecise and uncertain data have been introduced into databases for imperfect information processing by applying fuzzy logic, probability, and more generally soft computing.

It is crucial for databases to explicitly represent and process imprecise and uncertain data. This is because databases have been extensively applied in many application domains which may have a big deal of imprecision and vagueness. Imprecise and uncertain data can be found, for example, in the integration of data sources and data generation with nontraditional means (e.g., automatic information extraction and data acquirement by sensor and RFID).

Probabilistic theory can bridge the gap between human-understandable soft logic and machine-readable hard logic, and has been a crucial means of implementing intelligent data processing and intelligent information systems. In order to deal with probabilistic information in databases, currently the research and development of probabilistic data management are attracting an increased attention.

This book covers a fast-growing topic in great depth and focuses on the technologies and applications of probabilistic data management. It aims to provide a single account of current studies in probabilistic data management. The objective of the book is to provide the state of the art information to researchers, practitioners, and graduate students of information technology, and at the same time serving the information technology professional faced with non-traditional applications that make the application of conventional approaches difficult or impossible.

This book consists of six chapters. The first two chapters focus on probabilistic data management in the context of databases, discussing probabilistic spatiotemporal databases and probabilistic object-oriented databases with fuzzy measures, respectively. The next two chapters present probabilistic data management in the context of XML. The final two chapters focus on probabilistic data management in other frameworks, covering uncertain and imprecise multidimensional data streams in OLAP and tractable probabilistic description logic programs for the Semantic Web.

Chapter 1 focuses on research in probabilistic spatiotemporal databases and presents an overview on this topic. Particularly, for the results about probabilistic spatiotemporal databases using the SPOT (Spatial PrObabilistic Temporal) approach, this chapter provides a uniform overview. Also the chapter presents numerous interesting research problems using the SPOT framework for probabilistic spatiotemporal databases that await further work.

Chapter 2 concentrates on modeling probabilistic evens with fuzzy measures in the object-oriented database model. Instead of crisp probability measures or interval probability measures of objects and classes, fuzzy sets are applied to represent imprecise probability measures. A probabilistic object-oriented database model with fuzzy measures is introduced, which incorporates fuzzy probability degrees to handle imprecision and uncertainty. Based on the proposed probabilistic object-oriented database model, several major semantic relationships of objects and classes, including equivalent object relationships, object-class relationships and subclass-superclass relationships, are investigated.

Chapter 3 aims to model and manage various kinds of uncertain data in probabilistic XML. The chapter reviews the literature on probabilistic XML. Specifically, this chapter discusses the probabilistic XML models that have been proposed and the complexity of query evaluation therein. Also the chapter discusses other data-management tasks for probabilistic XML like updates and compression, as well as systemic and implementation aspects.

Chapter 4 surveys a few applications in sensor networks, ubiquitous computing, and scientific databases that require managing uncertain and probabilistic data. The chapter presents two approaches to meeting this requirement. The first approach is proposed for a rich treatment of probability distributions in the system, in particular the SPO framework and the SP-algebra. The second approach stays closer to a traditional DBMS, extended with tuple probabilities or attribute probability distributions, and studies the semantics and efficient processing of queries.

Chapter 5 introduces a novel approach for tackling the problem of OLAPing uncertain and imprecise multidimensional data streams via novel theoretical tools that exploit probability, possible-worlds and probabilistic-estimators theories. The result constitutes a fundamental study for this scientific field that, behind to elegant theories, is relevant for a plethora of modern data stream applications and systems that are more and more characterized by the presence of uncertainty and imprecision.

Chapter 6 proposes tractable probabilistic description logic programs (dl-programs) for the Semantic Web, which combine tractable description logics (DLs), normal programs under the answer set and the well-founded semantics, and probabilities. The chapter first provides novel reductions of tight query processing and of deciding consistency in probabilistic dl-programs under the answer set semantics to the answer set semantics of the underlying normal dl-programs. Based on these reductions, the chapter then introduces a novel well-founded semantics for probabilistic dl-programs, called the total well-founded semantics. The chapter presents an anytime algorithm for tight query processing in probabilistic dl-programs under the total well-founded semantics. It is also shown that tight literal query processing in probabilistic dl-programs under the total well-founded semantics can be done in polynomial time in the data complexity and is complete for EXP in the combined complexity. Finally, the chapter describes an application of probabilistic dl-programs in probabilistic data integration for the Semantic Web.

Acknowledgements

We wish to thank all of the authors for their insights and excellent contributions to this book and would like to acknowledge the help of all involved in the collation and review process of the book. Thanks go to all those who provided constructive and comprehensive reviews. Thanks go to Janusz Kacprzyk, the series editor of Studies in Fuzziness and Soft Computing, and Thomas Ditzinger, the senior editor of Applied Sciences and Engineering of Springer-Verlag, for their support in the preparation of this volume.

Northeastern University, China
February 2, 2013

Zongmin Ma and Li Yan

Contents

Research in Probabilistic Spatiotemporal Databases: The SPOT Framework

John Grant, Francesco Parisi, and V.S. Subrahmanian

Abstract. We start by providing an overview of research on probabilistic spatiotemporal databases. The bulk of the paper is a review of our previous results about probabilistic spatiotemporal databases using the SPOT approach. Presently these results are scattered in various papers and it is useful to provide a uniform overview. We also present numerous interesting research problems using the SPOT framework for probabilistic spatiotemporal databases that await further work.

1 Introduction

In recent years much interest has focused on the tracking of moving objects and reasoning about moving objects. Particularly with GPS systems it becomes possible to track vehicles, cell phones, supply items, RFID tags, and the importance of such tracking continues to increase. Clearly, space and time are important factors in this endeavor but probability is also useful. The reason for that is that the locations, and possibly the identity of the objects and the time may be known only with some uncertainty and we can express such uncertainty by the use of probability. Furthermore, we claim that in many cases the probability itself is not known exactly, in which case the use of a probability interval is appropriate. Thus our framework involves space, time, and probability intervals.

John Grant
Towson University, Towson, Maryland, USA, and
University of Maryland, College Park, Maryland, USA
e-mail: jgrant@towson.edu

Francesco Parisi
Università della Calabria, Rende (CS), Italy
e-mail: fparisi@deis.unical.it

V.S. Subrahmanian
University of Maryland, College Park, Maryland, USA
e-mail: vs@cs.umd.edu

Z. Ma & L. Yan (Eds.): *Advances in Probabilistic Databases*, STUDFUZZ 304, pp. 1–22.
DOI: 10.1007/978-3-642-37509-5_1

A couple of years ago we became interested in providing a formalism for the representation, querying, and updating of probabilistic spatiotemporal databases in a straightforward manner with a simple syntax and an intuitive model-theoretic semantics. For this purpose we introduced the SPOT (Spatial PrObabilistic Temporal) framework in which a basic statement (atomic formula) states that a particular object is in a particular region (for location) at a particular time with a probability that is in a particular probability interval.

We then implemented the SPOT formalism and tackled several issues concerning SPOT databases. In particular, we showed that several problems involving SPOT databases can be transformed into linear programming problems. We developed indexing methods to speed several types of selection queries. We also studied updating SPOT databases in the spirit of AGM-style belief revision. Altogether we published five papers on the SPOT approach. The purpose of this paper is to summarize our results in one place and to propose numerous research questions for the further study of probabilistic spatiotemporal databases in the SPOT framework.

We start in Section 2 by reviewing some of the important work done by other researchers related to probabilistic spatiotemporal databases that did not use the SPOT approach. Then Section 3 introduces the syntax and semantics of SPOT databases and provides an example that will be used throughout the paper to illustrate various concepts. Section 4 contains the fundamental results about SPOT databases including the transformation to linear programming. The processing of certain kinds of selection queries is discussed in Section 5; both optimistic and cautious answers are considered. Section 6 reviews the algorithms for database updating/revision. Finally, Section 7 contains interesting research questions.

2 Review of Previous Related Research

In order to place the SPOT framework in the proper context we now review research related to it. This is by no means an exhaustive survey; we give a historical background and briefly review some important papers; many additional references appear in the works we mention. The problem of predicting where moving objects will be in the future, when they will be there, and with what probability is intrinsically challenging. This can be easily understood by observing that reasoning with just one or two of these three aspects (probability, space and time) is already quite challenging. In fact, researchers have investigated separately probabilistic databases, probabilistic spatial databases, probabilistic temporal databases, and probabilistic spatiotemporal databases.

Kiessling and his group [28] develop the DUCK framework for reasoning with uncertainty. They provide an elegant, logical, axiomatic theory for uncertain reasoning in the presence of rules. In the same spirit as Kiessling et al., Ng and Subrahmanian [45] present a probabilistic semantics for deductive databases — they assume absolute ignorance, and furthermore, assume that rules are present in the system. Lakshmanan and Sadri [38] show how selected probabilistic strategies can be used to extend the previous probabilistic models. Lakshmanan and Shiri [39] demonstrate how deductive databases may be parameterized through the use of conjunction and disjunction strategies. Barbara et al. [3] develop a probabilistic data model and propose probabilistic operators. Their work is based on the assumption that probabilities of compound events can always be precisely determined. Cavallo and Pittarelli [8, 54], propose a model for probabilistic relational databases in which tuples in a probabilistic relation are interpreted using an exclusive or. Dey and Sarkar [15] propose an elegant `1NF` approach to handling probabilistic databases. In another work Kifer and Li [33] examine quantitative logic programming and introduce formal semantics for such systems. Other systems from the probabilistic database community also provide insight into probabilistic information reasoning and storage [18, 19, 13, 34, 31]. Lukasiewicz and his colleagues [41, 42] study probabilistic reasoning in logic programming, as does Dekhtyar [14]. However, none of these works explicitly handle space or time.

The problem of efficiently storing and querying data representing spatial predictions has deserved much attention by many researchers. Tao et al [60] develop an indexing structure for spatial probabilistic data designed to solve a specific problem. They assume that there is a single probability distribution function detailing where an object might be at a given point in time in the entire space and their focus is on optimizing access to that probability density function. Using an R-tree-inspired U-tree indexing structure they use hyperplanes to approximate the evolution of these probabilistically constrained regions between time points. In [46], methods for dealing with positionally uncertain spatial data are considered. Their data model associates each point with a cluster, where points in the same cluster have the same error. This model also allows only one pdf. The authors describe a PrR-tree for storing and querying survey data, which uses a rectangular bounding region whose corners are defined via Gaussian distributions. Like the above work, in [40], Lian et al use a data model with one pdf over each object's locations. They introduce probabilistic group nearest neighbor queries, where given a set of points and a probability threshold the system returns the set of objects that have minimal aggregate distance to the set of points with a probability over the threshold. Dai et al [12] focus on probabilities for the *existence* of a given object at a given point without worrying about the possibility of the object being at another point. They show how to build an augmented R-tree and use that tree to answer selection queries more effectively than considering probability as an extra dimension in the R-tree. [7] uses a paradigm called "line simplification" to approximate trajectories of moving objects, though their framework does not involve uncertainty.

On the purely temporal side, Snodgrass is one of the first to model indeterminate instances [59] — he proposes the use of a model based on three valued logic. Dutta [17] later give a fuzzy logic based approach to handle *generalized temporal events* — events that may occur multiple times. This approach is also used by Dubois and Prade [16]. Gadia [25] proposes an elegant model to handle incomplete temporal information as well. He models values that are completely known, values that are unknown but are known to have occurred, values that are known if they have occurred, and values that are unknown even if they occurred. Koubarakis [36] proposes the use of constraints for representing temporal data. Brusoni [6] develops a system called LaTeR that restricts constraints to conjunctions of linear inequalities, as does Koubarakis' work. [4] develops a framework to track uncertainty, time, and the pedigree of data, but does not handle spatial information. As a matter of fact none of these works handle both space and time.

There is also much work on *spatio-temporal logics* [24, 43] in the literature. These logics extend temporal logics to handle space. This includes work on qualitative spatio-temporal theories (for a survey see [11, 44] [58] which discusses the frame problem when constructing a logic-based calculus for reasoning about the movement of objects in a real-valued co-ordinate system). [55] focuses on relative position and the orientation of objects with existing methods for qualitative reasoning in a Newtonian framework. Other efforts combine a spatial logic, such as $RCC-8$ [56], $BRCC-8$ [62] and $S4_u$ [5], with propositional temporal logics ($\mathscr{PTL}$). The work on spatio-temporal reasoning is mostly qualitative [11, 43, 63, 23], and focuses on relations between spatio-temporal entities while dealing with discrete time. However, these works are not intended for reasoning about moving objects whose location at a given point in time (past, present or future) is uncertain (they not consider probabilities).

In addition to the above works on spatio-temporal logics, there are works on logics integrating time and probabilities. Much of this work was performed in the model checking community. The PRISM system [37] supports a mix of time and probabilities for model checking with respect to specifications written in the temporal probabilistic logics PCTL [30] and CSL [2]. However, none of these works has any spatial element in them, and they focus on model checking, not on handling knowledge bases. The work on "go" theories [20, 22, 21] focuses on spatio-temporal logical theories that are sets of "go" atoms. Such atoms intuitively describe known plans of moving objects. A go-atom states that an object will go from location A to location B, leaving A at a time point in some time interval, arriving at B at a time point in some interval, and traveling in the interim at a velocity within some stated interval. [20] develops a basic theory of "go" theories, while [22] gives a closed world assumption for such theories. Later, [51] extends this logic to include some probabilistic information about such plans. Finally, the SPOT framework extends this work to uncertainty about where objects might be at a given time [50, 49].

While there is substantial work in indexing spatial temporal data without probabilities [52, 61, 35, 1, 53, 29], none of these works address a data model compatible with our SPOT framework: they suppose no probabilities and model object movement as linear. SPOT databases were developed by the authors in past work [50, 49] to store such predictions without making the assumptions in prior work[60, 12, 7, 4]. In fact, our observations about the advantages of the SPOT approach hold also for some very recent papers we mention in the next paragraph.

Up to this point we reviewed work done prior to our formulation of SPOT databases. We end this section with a brief review of some recent papers on probabilistic spatiotemporal databases. These papers do not use the SPOT approach. Chung et al. [10] derive a pdf for the location of an object moving in a one-dimensional space by using its past moving behavior or the moving velocity distribution. Their probabilistic range queries find objects that are in a specified region of space within a specified time interval, and with a probability that is at least a threshhold value. Their indexing takes care of eliminating object that are too far from consideration. Zhang et al. [65] provide a framework that allows their model to be incorporated easily into existing DBMSs and work for all objects even if their location and velocity are uncertain and the movements are unusual. For indexing they use a B^x-tree which is a variant of a B^+-tree that is applicable to moving objects. Yang et al. [64] work with moving objects in indoor space which is different from Euclidean and spatial network spaces. Their query asks for all sets of k objects that have at least a threshold probability of containing the k nearest objects to a given object. This paper defines a minimal indoor walking distance and uses two types of pruning to efficiently solve such queries. Chen et al. [9] deal with a similar problem but use a TPR-tree for indexing. They also deal with the query result quality by using both a false positive and a false negative rate. Finally, the most recent paper we mention, [66] deals primarily with objects moving along road networks, certainly an important application. They introduce a novel indexing mechanism called UTH (Uncertain Trajectories Hierarchy) for efficiently processing probabilistic range queries.

3 A SPOT Database

This section reviews the syntax and semantics of SPOT databases given in [50].

Syntax

We assume the existence of a set ID of objects ids, a set T of time points ranging over the integers, and a finite set $Space$ of points. Unless stated otherwise, we assume that $Space$ is a grid of size $N \times N$ where we only consider integer coordinates [1]. We assume that an object can be in only one location at a time, but that a single location may contain more than one object.

A *rectangle* is any region in $Space$ that can be described by constraints of the form $left \le x \le right$ and $bottom \le y \le top$ where $left, right, bottom, top$ are integers in $[0..N]$. Thus, all rectangles have edges parallel to the x and y-axes. A rectangle is empty if either $left > right$ or $bottom > top$.

Definition 1 (**SPOT atom/database**). A SPOT *atom* is a tuple $(id, r, t, [\ell, u])$, where $id \in ID$ is an object id, $r \subseteq Space$ is a non-empty rectangular region in the space, $t \in T$ is a time point, and $\ell, u \in [0,1]$ are probability bounds with $\ell \le u$. A SPOT *database* is a finite set of SPOT atoms.

Intuitively, the SPOT atom $(id, r, t, [\ell, u])$ says that objects id is/was inside the specified region r at time t with probability in the $[\ell, u]$ interval.

[1] The framework is easily extensible to higher dimensions.

Example 1. Consider a lab where data coming from biometric sensors are collected, analyzed and stored. Biometric data such as faces, voices, and fingerprints recognized by sensors are matched against given profiles (such as those of people having access to the lab) and tuples like those in Table 1 are obtained. Every tuple consists of the profile id resulting from the matching phase, the area of the lab where the sensor recognizing the profile is operating, the time point at which the profile has been recognized, and the lower and upper probability bounds of the recognizing process getting the tuple. For instance, the tuple in the first row of Table 1 representing the SPOT atom $(id_1, d, 1, [0.9, 1])$ says that profile having id id_1 was in region d at time 1 with probability in the interval $[0.9, 1]$. In Figure 1, the plan of the lab and the areas covered by biometric sensors are shown. In area d a fingerprint sensor is located, whose high accuracy entails a narrow probability interval with upper bound equal to 1. After fingerprint authentication, the id_1 profile was recognized at time 3 in areas b and c with probability in $[0.6, 1]$ and $[0.7, 0, 8]$, respectively.

Table 1 SPOT database $\mathcal{S}_{lab}$

Id	Area	Time	Lower Prob	Upper Prob
id_1	d	1	0.9	1
id_1	b	3	0.6	1
id_1	c	3	0.7	0.8
id_2	b	1	0.5	0.9
id_2	e	2	0.2	0.5
id_2	e	3	0.6	0.9

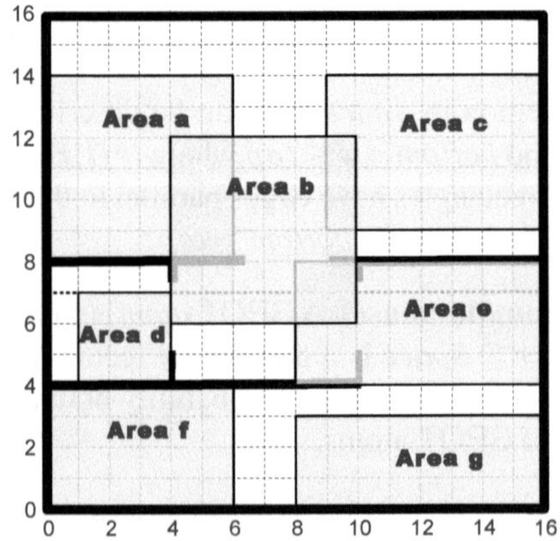

Fig. 1 Areas of the lab

Given a SPOT database $\mathscr{S}$, a fixed object id and a fixed time t, we use the notation $\mathscr{S}^{id,t}$ to refer to the set:

$$\mathscr{S}^{id,t} = \{(id',r',t',[\ell',u']) \in \mathscr{S} \mid id' = id \wedge t' = t\}.$$

Semantics

The meaning of a SPOT database is given by the set of interpretations that satisfy it.

Definition 2 (SPOT interpretation). A SPOT *interpretation* is a function $I : ID \times Space \times T \to [0,1]$ such that for each $id \in ID$ and $t \in T$,

$$\sum_{p \in Space} I(id,p,t) = 1.$$

For a given an interpretation I, we sometimes abuse notation and write $I^{id,t}(p) = I(id,p,t)$. In this case, $I^{id,t}$ is a probability distribution function (PDF).

Example 2. Interpretation I_1 for the SPOT database $\mathscr{S}_{lab}$ introduced in Example 1 is as follows.

$$\begin{array}{ll}
I_1(id_1,(3,6),1) = 0.4 & I_1(id_1,(2,5),1) = 0.2 \\
I_1(id_1,(3,5),1) = 0.3 & I_1(id_1,(5,5),1) = 0.1 \\
I_1(id_1,(7,5),2) = 0.5 & I_1(id_1,(4,2),2) = 0.5 \\
I_1(id_1,(10,10),3) = 0.7 & I_1(id_1,(7,5),3) = 0.3 \\
I_1(id_2,(8,10),1) = 0.1 & I_1(id_2,(12,12),1) = 0.9 \\
I_1(id_2,(9,7),2) = 0.3 & I_1(id_2,(12,13),2) = 0.7 \\
I_1(id_2,(14,5),3) = 0.8 & I_1(id_2,(12,14),3) = 0.2
\end{array}$$

Moreover, $I_1(id,p,t) = 0$ for all triplets (id,p,t) not mentioned above.

Given an interpretation I and region r, the probability that object id is in r at time t *according to I* is $\Sigma_{p \in r} I(id,p,t)$. We now define satisfaction by an interpretation.

Definition 3 (Satisfaction). Let $a = (id,r,t,[\ell,u])$ be a SPOT atom and let I be a SPOT interpretation. We say that I satisfies a (denoted $I \models a$) iff $\Sigma_{p \in r} I(id,p,t) \in [\ell,u]$. I satisfies SPOT database $\mathscr{S}$ (denoted $I \models \mathscr{S}$) iff I satisfies every atom in $\mathscr{S}$.

Example 3. Continuing with our running example, interpretation I_1 satisfies the SPOT atom $(id_1,d,1,[0.9,1])$ as, for id id_1 and time point 1, I_1 assigns probability 0.4 to $(3,6)$, 0.3 to $(3,5)$, and 0.2 to $(2,5)$ (which are points in area d), and probability 0.1 to $(5,5)$ which is outside area d. Hence, the probability that id_1 is in area d at time point 1 is 0.9, which belongs to the interval $[0.9,1]$ specified by the considered SPOT atom. Reasoning analogously, it is easy to see that I_1 satisfies all of the atoms in Table 1 except for $(id_2,b,1,[0.5,0.9])$, as the probability to be in area b at time 1 for id id_2 is set to 0.1 by I_1, instead of a value in $[0.5,0.9]$.

Let I_2 be an interpretation equal to I_1 except that $I_2(id_2,(8,10),1) = 0.7$ and $I_2(id_2,(12,12),1) = 0.3$. This interpretation satisfies the SPOT database $\mathscr{S}_{lab}$.

4 The Basic Definitions and Results

We use $\mathbf{I}(\mathcal{S})$ to denote the set of interpretations that satisfy a SPOT database $\mathcal{S}$, that is, $\mathbf{I}(\mathcal{S}) = \{I \mid I \models \mathcal{S}\}$.

Definition 4 (Consistency). A SPOT database $\mathcal{S}$ is *consistent* iff $\mathbf{I}(\mathcal{S}) \neq \emptyset$.

Definition 5 (Compatibility). A SPOT atom a is compatible with a SPOT database $\mathcal{S}$, denoted as $a \Subset \mathcal{S}$, iff $\mathcal{S} \cup \{a\}$ is consistent.

Definition 6 (Entailment). A SPOT database $\mathcal{S}$ *entails* a SPOT atom a, denoted as $\mathcal{S} \models a$, iff $\forall I \in \mathbf{I}(\mathcal{S})$, $I \models a$. A SPOT database $\mathcal{S}_1$ entails a SPOT database $\mathcal{S}_2$, denoted as $\mathcal{S}_1 \models \mathcal{S}_2$, iff $\forall a \in \mathcal{S}_2$, $\mathcal{S}_1 \models a$.

Example 4. Interpretation I_2 of Example 3 proves that the SPOT database $\mathcal{S}_{lab}$ of Example 1 is consistent. It is easy to see that $a = (id_1, f, 2, [0, 0.5]) \Subset \mathcal{S}_{lab}$ because $I_2 \models \mathcal{S}_{lab} \cup \{a\}$ and $\mathcal{S}_{lab} \models (id_1, d, 1, [0.75, 1])$.

Given a SPOT database $\mathcal{S}$, an object $id \in ID$, and a time point $t \in T$, that is, $\mathcal{S}^{id,t}$, [50] defined a set $LC(\mathcal{S}, id, t)$ of linear constraints. $LC(\mathcal{S}, id, t)$ uses variables v_p to denote the probability that object id will be at point $p \in Space$ at time t.

Definition 7 ($LC(\cdot)$)**.** For SPOT database $\mathcal{S}$, $id \in ID$, and $t \in T$, $LC(\mathcal{S}, id, t)$ contains:

- $\forall (id, r, t, [\ell, u]) \in \mathcal{S}^{id,t}$,
  $\left(\sum_{p \in r} v_p \geq \ell\right) \in LC(\mathcal{S}, id, t)$,
  $\left(\sum_{p \in r} v_p \leq u\right) \in LC(\mathcal{S}, id, t)$
- $\left(\sum_{p \in Space} v_p = 1\right) \in LC(\mathcal{S}, id, t)$,
- $\forall p \in Space \quad (v_p \geq 0) \in LC(\mathcal{S}, id, t)$
- No other constraints are in $LC(\mathcal{S}, id, t)$.

The problem of checking the consistency of a SPOT database was addressed in [50], where it was shown that SPOT database $\mathcal{S}$ is consistent iff $LC(\mathcal{S}, id, t)$ is feasible for all $\langle id, t \rangle$ pairs. The compatibility and entailment of a SPOT atom can be checked via the following result shown in [50].

Theorem 1. *Given a* SPOT *database* $\mathcal{S}$ *and a* SPOT *atom* $(id, r, t, [\ell, u])$,

i) $(id, r, t, [\ell, u]) \Subset \mathcal{S}$ *iff* $LC(\mathcal{S} \cup \{(id, r, t, [\ell, u])\}, id, t)$ *is feasible.*
ii) $\mathcal{S} \models (id, r, t, [\ell, u])$ *iff* $[\ell', u'] \subseteq [\ell, u]$ *where*
- $\ell' =$ ***minimize*** $\sum_{p \in r} v_p$ ***subject to*** $LC(\mathcal{S}, id, t)$
- $u' =$ ***maximize*** $\sum_{p \in r} v_p$ ***subject to*** $LC(\mathcal{S}, id, t)$.

Hence, the consistency of a SPOT database, as well as the compatibility and entailment of a SPOT atom, can be checked by a linear programming algorithm. The complexity of these algorithms was shown to be $O(|ID| \cdot |T| \cdot (|Space| \cdot |\mathcal{S}|)^3)$.

An optimized version of these algorithms was proposed in [49], where the number of variables of $LC(\cdot)$ was drastically reduced by introducing an equivalence relation on points in *Space*, giving a partition of *Space*, $\mathscr{P} = \{\mathscr{P}_1, \ldots, \mathscr{P}_n\}$ so that all the points in any $\mathscr{P}_i$ cannot be distinguished by $\mathscr{S}^{id,t}$. It was proved that consistency, compatibility, and entailment can be checked by considering a version of $LC(\cdot)$ (namely $PLC(\cdot)$) where the points in *Space* are replaced by the partitions in $\mathscr{P}$, so that the points $p \in \mathscr{P}_i$ are replaced by a new single variable $v_{\mathscr{P}_i}$. As the size of $\mathscr{P}$ is typically very much lower than the number of points in *Space*, the amount of time needed for solving $PLC(\cdot)$ is very much smaller than that needed for solving $LC(\cdot)$ [49] experimentally shows that reduced-size algorithms are much more efficient than those introduced in [50].

5 Query Processing in SPOT

The most investigated kind of query in SPOT database is selection.

Definition 8 (Selection query). A *selection query* is an expression of the form $(?id, q, ?t, [\ell, u])$ where q is a region of *Space*, not necessarily rectangular, $[\ell, u]$ is a probability interval, $?id$ is a variable ranging over ids in ID, and $?t$ is a variable ranging over time points in T.

Intuitively, a selection query says: "Find all objects *id* and times t such that the object *id* is inside the specified region q at time t with a probability in the $[\ell, u]$ interval." There are two semantics for interpreting this statement, leading to two types of selection queries.

Definition 9 (Optimistic/Cautious selection). Suppose $\mathscr{S}$ is a SPOT database and $(?id, q, ?t, [\ell, u])$ is a selection query.
The *optimistic answer* to $(?id, q, ?t, [\ell, u])$ is the set

$$\{\langle id, t\rangle \mid id \in ID \wedge t \in T \wedge (id, q, t, [\ell, u]) \Cap \mathscr{S}\}.$$

The *cautious answer* to $(?id, q, ?t, [\ell, u])$ is the set

$$\{\langle id, t\rangle \mid id \in ID \wedge t \in T \wedge \mathscr{S} \models (id, q, t, [\ell, u])\}.$$

Optimistic selection returns objects and time points that *may* be in the query region with probability in the specified interval, whereas cautious selection only returns those objects and time points that are *guaranteed* to be in that region with probability in that interval. Thus, the cautious answer is a subset of the optimistic one.

Example 5. Continuing our running example, one may be interested in knowing the ids and time points of profiles that were in the room where the fingerprint sensor is located, with probability greater than 0.75. This can be expressed by the selection query $(?id, q, ?t, [0.75, 1])$, where q is the rectangle defined by constraints $0 \le x \le 6$ and $4 \le y \le 8$ (this query region includes the whole area d, a portion of area b, and some other points). The optimistic answer of this query is the set $\{< id_1, 1 >, < id_1, 2 >, < id_1, 3 >, < id_2, 1 >\}$, whereas the cautious answer only contains the pair $< id_1, 1 >$.

Optimistic and cautious selection can be computed by exploiting the results of Theorem 1. Specifically, given the selection query $Q = (?id, q, ?t, [\ell, u])$ over the SPOT database $\mathscr{S}$, the optimistic answer to Q can be computed by solving, for each pair $\langle id, t\rangle$ in $\mathscr{S}$, the linear program $LC(\mathscr{S} \cup \{(id, q, t, [\ell, u])\}, id, t)$: if it has a solution then $\langle id, t\rangle$ is in the optimistic answer of Q. The cautious answer to Q can be computed by solving, for each pair $\langle id, t\rangle$ in $\mathscr{S}$, the two optimization problems of Theorem 1(ii) which return the interval $[\ell', u']$. Then, checking if $[\ell', u'] \subseteq [\ell, u]$ is sufficient for deciding if the pair $\langle id, t\rangle$ is in the cautious answer of Q. We refer to these approaches as the *naive* algorithms for optimistic and cautious selection.

Efficient algorithms for computing optimistic and cautious selection were proposed in [49] and [47], respectively. Both proposed approaches exploit strategies for pruning the search space of candidate answers of a given query.

Optimistic Selection

An index structure, called SPOT -tree, and algorithms to compute optimistic answers to selection queries using the index were proposed in [49]. Each node of a SPOT -tree is labeled with a *composite* SPOT atom, which compactly represents a set of SPOT atoms (that is, a SPOT database). The relationship between parent and children nodes of a SPOT -tree is based on logical implication of SPOT databases. Basically, there is an entailment relationship between each composite atom labeling a child node and the composite atom labeling its parent node. Hence, the composite atom labeling the root node of a SPOT -tree is entailed by every composite atom labeling any node of the tree. Further, composite atoms labeling leaf nodes are entailed by SPOT atoms in the database.

The SPOT -tree reduces the set of $< id, t >$ pairs to be considered as candidates for the optimistic answer of a selection query — potentially, this set contains all the $< id, t >$ pairs in the SPOT database. The logical relationship between nodes of a SPOT -tree entails that, for each child node, the set of SPOT interpretations of its composite atom is a subset of the set of SPOT interpretations of its parent node composite atom. As a consequence, given a selection query Q, if the composite atom labeling the node n is not compatible with Q (this condition can be checked in constant time due to the structure of composite atoms), then any composite atom of the subtree rooted in n is not compatible with Q, which in turns means that all the $< id, t >$ pairs of SPOT -tree rooted in n do not belong to the optimistic answer of Q, i.e., they can be pruned from the search space.

For $< id, t >$ pairs that cannot be pruned by traversing the SPOT -tree, the naive algorithm is used for checking whether they are in the optimistic answer. Thus, for every pair pruned, we save the time needed for solving the linear programming problem of Theorem 1(i), as confirmed by the experimental study in [49].

Cautious Selection

In [47] the problem of efficiently computing cautious answers to selection queries was investigated. The proposed approach is based on geometric considerations which follow from the fact that both SPOT databases and selection queries define convex polytopes in the so-called SPOT PFD Space. The relationship between these polytopes can be used to answer queries. Specifically, it was shown that an $< id,t >$ pair belongs to the cautious answer of a given selection query iff the polytope defined by the query contains that defined by the SPOT database $\mathscr{S}^{id,t}$.

Approximations of the $\mathscr{S}^{id,t}$ polytope by interior and containing regions were introduced, instead of using the original $\mathscr{S}^{id,t}$ polytope which would have lead to an approach as computationally expensive as solving optimization problems of Theorem 1(ii). Thus, an $< id,t >$ pair is in the cautious answer of a given selection query if the query polytope contains a region containing the $\mathscr{S}^{id,t}$ polytope (this ensures that the query polytope contains that defined by $\mathscr{S}^{id,t}$). Similarly, an $< id,t >$ pair in *not* in the cautious answer (i.e., it can be pruned) if the query polytope does not contain an interior region of the $\mathscr{S}^{id,t}$ polytope. Both of these strategies can be jointly used to prune the search space when answering a query.

Efficient ways of finding interior and containing regions were also proposed. Containing regions can be obtained by starting from composite atoms introduced in [49] for defining SPOT -tree nodes. Internal regions can be obtained by following an inline or preprocessing approach. The former consist of storing solutions (that is, internal points of $\mathscr{S}^{id,t}$ polytope) of previously asked selection queries and then building the convex envelopes of found points. The latter approach, which is preferable to inlining when spare resources are available for precomputation, consists of solving a few optimization problems to find some internal points to be used to construct their convex envelope. As experimentally proved in [47], using either interior and containing regions yields improvement in terms of the efficiency of cautious selection.

SPOT *Query Algebra*

A relational-style algebra for SPOT databases has been proposed in [50]. This algebra contains a version of the selection, union, intersection, difference and join operators. Specifically, selection considered in [50] is a special case of cautious selection which returns ids only. Processing such queries by linear programming algorithms can be avoided if the database is known to satisfy the following *disjointness* property: there are no two distinct SPOT atoms in the database referring to the same object, the same time point and intersecting regions of *Space*.

Set operations and join in SPOT databases are more complicated and challenging than selection. We start by discussing the union operator, the simplest one. Union adds restrictions on the set of SPOT interpretations of the databases involved in the union operation. The intuition is that adding new information to a SPOT database reduces the degree of freedom that one has in choosing an interpretation. Formally, the union of SPOT databases $\mathscr{S}_1$ and $\mathscr{S}_2$ results in a SPOT database whose set of interpretation is $\mathbf{I}(\mathscr{S}_1) \cap \mathbf{I}(\mathscr{S}_2)$. On the contrary, intersection and difference remove restrictions on SPOT interpretations of databases involved. Semantically, one would expect that $\mathbf{I}(\mathscr{S}_1 \cap \mathscr{S}_2) = \mathbf{I}(\mathscr{S}_1) \cup \mathbf{I}(\mathscr{S}_2)$ and $\mathbf{I}(\mathscr{S}_1 - \mathscr{S}_2) = \mathbf{I}(\mathscr{S}_1) \setminus \bar{\mathbf{I}}(\mathscr{S}_2)$, where $\bar{\mathbf{I}}(\mathscr{S}_2) = \{I \mid I \notin \mathbf{I}(\mathscr{S}_2)\}$. However, such semantics for intersection and difference cannot be expressed by SPOT databases since disjunction of probability intervals of SPOT atoms would be required. Hence, weaker definitions of intersection and difference operators returning supersets of the expected sets of interpretations (defined above) were introduced. Finally, a version of join was introduced that puts together information from different SPOT atoms having the same $< id, t >$ pair and satisfying an input function combining atom regions.

Nearest Neighbor Queries

Another kind of query in which SPOT database users may be interested is a nearest neighbor query, that is finding the nearest expected object neighbor to a given point of *Space* at a given time. As an example, a cell phone company may want to know the nearest cell phone to a given cell tower. Also the nearest neighbor to a given object, whose position is not known exactly, at a given time may be needed, such as finding the cell phone nearest to another one at a given time. Dealing with nearest neighbor queries implicitly requires dealing with the expected distances of objects from a given point or object at a given time. Here the expression "expected distance" is used in the strict sense of statistical expected values.

As SPOT databases do not state exactly where objects are at any given time, nor the exact probability that an object is in a given region of *Space* at a given time, minimal expected distance and maximal expected distance between objects were investigated in [50]. Then, these concepts were used to define two versions of nearest expected neighbor queries which return the object id whose expected distance from a given point at a given time is minimum or maximum, respectively. Under the assumption that the database is disjoint, algorithms for computing expected distance and nearest neighbor queries were proposed and experimentally validated.

6 Updates in SPOT

As SPOT databases provide information on moving objects, one of the aspect to be addressed is that information on these objects may continuously change as objects move. An object may encounter unexpected situations during its move, which may lead to a revision of estimates of where it may be in the future, as well as a revision of where it may have been in the past. The problem of revising SPOT data was first addressed in [48] and then further investigated in [27], where several strategies for revising SPOT data were proposed.

If the insertion of a SPOT atom a into a SPOT database $\mathscr{S}$ leads to no inconsistency then the atom can just be added to the database. However, when it leads to inconsistency, that is $\mathbf{I}(\mathscr{S} \cup \{a\}) = \emptyset$, then many different belief revision operations are possible. A first revision strategy proposed in [27] consists of finding a maximal (w.r.t. either subset inclusion or cardinality) subset $\mathcal{S}'$ of $\mathcal{S}$ such that $\mathcal{S}' \cup \{a\}$ is consistent.

Example 6. Assume that the SPOT atom $a = (id_1, e, 3, [0.5, 1])$ providing new fresh information on the profile id_1 recognized at time 3 by the sensor in area e should be added to the SPOT database $\mathcal{S}_{lab}$ of Example 1. It is easy to see that $\mathcal{S}_{lab} \cup \{a\}$ is not consistent, as id_1 cannot stay at the same time in the disjoint areas c and e with probability in the intervals $[0.7, 0.8]$ and $[0.5, 1]$, respectively. A maximal consistent subset revision of $\mathcal{S}_{lab}$ is $\mathcal{S}'_{lab} = \mathcal{S}_{lab} \setminus \{(id_1, c, 3, [0.7, 0.8])\}$. As this revision removes only one atom from the database, it is maximal under both the subset inclusion and the subset cardinality criterion.

Maximal consistent subset revision works at the tuple level in the sense that the basic primitive to update the SPOT database is deletion of whole tuples. Strategies exploiting finer basic primitives such as updates of attribute values of tuples are also possible. For instance, in the example above, a revision may consist of changing the value of the temporal component of the tuple $(id_1, c, 3, [0.7, 0.8])$ in $\mathcal{S}_{lab}$ from 3 to 2. Revising strategies consisting of minimally modifying the spatial, temporal, or object components in $\mathcal{S}$ were investigated in [27]. Further, three revision mechanisms based on minimally revising the probability intervals in a SPOT database were addressed. In particular, by changing the relevant atoms' probability intervals to $[0, 1]$, consistency is restored; however, such a change loses much information and will usually not be minimal. It turns out that all of above-cited revision mechanisms lead to computational intractability except the strategy that revises the probability bounds. Probability revision can be computed by solving a linear programming problem similar to $LC(\mathcal{S}, id, t)$ (see Definition 7) where, for each SPOT atom a_i, additional variables low_i and up_i are used to represent the atom's modified lower and upper probability bounds.

Example 7. A probability revision for the SPOT database $\mathcal{S}_{lab}$ w.r.t. atom $a = (id_1, e, 3, [0.5, 1])$ is obtained by solving the following linear programming problem which is obtained by considering that $\mathcal{S}_{lab}^{id_1,3} = \{(id_1, b, 3, [0.6, 1]), (id_1, c, 3, [0.7, 0.8])\}$:

$$\textbf{\textit{minimize}}\ (0.6 - low_1) + (up_1 - 1) + (0.7 - low_2) + (up_2 - 0.8)$$
$$\textbf{\textit{subject to}}$$
$$\begin{cases} low_1 \leq \sum_{p \in b} v_p \leq up_1 \\ 0 \leq low_1 \leq 0.6 \\ 1 \leq up_1 \leq 1 \\ low_2 \leq \sum_{p \in c} v_p \leq up_2 \\ 0 \leq low_2 \leq 0.7 \\ 0.8 \leq up_2 \leq 1 \\ 0.5 \leq \sum_{p \in e} v_p \leq 1 \\ \sum_{p \in Space} v_p = 1 \\ \forall p \in Space \quad v_p \geq 0 \end{cases}$$

Basically, for each atom $a_i = (id, r_i, t, [\ell_i, u_i])$ in the database, where id and t are the object identifier and the time point of the atom added to the database, the modified probability interval $[low_i, up_i]$ is such that $[\ell_i, u_i] \subseteq [low_i, up_i] \subseteq [0,1]$. The inequalities with variables v_p, where p is a point in *Space*, ensure that there is a PDF satisfying the revision atom and every atom a_i in the database with $[\ell_i, u_i]$ replaced by $[low_i, up_i]$. Finally, the objective function guarantees that the revised probability bounds are as close as possible to the original ones.

In [48] and [27], the SPOT framework was extended to remove the assumptions that there are no velocity constraints on moving objects and that all points in $\mathscr{S}$ are reachable from all other points by all objects. A general notion of *reachability definition* was provided to capture the scenario where objects have speed limits and only some points are reachable by objects depending on both the distance between the points, the objects' speed, and possible obstacles in the way. Reachability constraints require the use of more complex variables in the linear programming problem of Definition 7. However, in this case the size of the problem increases quadratically in the size of *Space* and the maximum time range in $\mathscr{S}$. As exact optimizations like that discussed after Theorem 1 do not work in this case due to the new structure of the linear programming problem, a suite of heuristics was proposed in [27] in order to speed up the computation of probability revision.

7 Research Questions about SPOT

As sketched in the previous sections we have found SPOT to be a very useful approach to the study of probabilistic spatiotemporal databases. We believe that there are interesting research problems within the SPOT framework as well through various extensions of SPOT . The purpose of this section is to specify and elaborate on such issues, thereby inviting researchers to do further work in this area. The topics are mostly independent of each other, although some could be combined.

We present five general topics. The first two are within the present SPOT framework; the other three present extensions to SPOT . First we discuss additional types of selection queries not considered in our papers. Then we suggest investigating the handling of aggregates and views. Recalling that SPOT databases use spatial regions and probability intervals, we suggest an extension to SPOT that allows for groups of objects and temporal intervals. Another possible extension involves adding connectives and quantifiers, thereby creating a full SPOT logic. Finally we suggest combining SPOT with logical terminology involving space and time.

7.1 Additional Selection Queries

As we discussed in Section 4, in our papers we dealt with selection queries of the form $(?id, q, ?t, [\ell, u])$ finding pairs of $\langle id, time \rangle$ values that provide answers to the query for a specific region q and probability interval $[\ell, u]$ using both the optimistic and cautious semantics. For the purpose of this section we will write such a query as $\{\langle id, t \rangle \mid (id, q_c, t, [\ell_c, u_c])\}$ indicating by the use of the subscript c that the region and probability values are fixed constants. We dealt with this selection query because we thought it was important, and we devised index structures for its efficient processing.

We believe that these and other index structures will be useful in answering other types of selection queries. In this subsection we list some queries with a similar form.

Let us start with atomic queries. Considering these systematically we start with one variable queries where the answer will be a set of values. Two of these appear interesting, namely $\{\langle id\rangle \mid (id, q_c, t_c, [\ell_c, u_c])\}$ and $\{\langle t\rangle \mid (id_c, q_c, t, [\ell_c, u_c])\}$. These are just like the queries we considered previously, but simpler, because in the first case we are considering a single time value and in the second case a single id value. The other two one variable queries appear less interesting because in the case of $\{\langle q\rangle \mid (id_c, q, t_c, [\ell_c, u_c])\}$ there may be many regions in the answer that are almost the same; while for $\{\langle [\ell, u]\rangle \mid (id_c, q_c, t_c, [\ell, u])\}$ the probability interval $[0,1]$ is always in the answer. In the latter case we would likely be interested only in the smallest such probability interval. So, for example, if the answer is $[.8,1]$, that means that id_c is in the region q_c at time t_c with probability at least .8.

We move on to two variable queries. Altogether there are 6 two variable queries. The query $\{\langle id, t\rangle \mid (id, q_c, t, [\ell_c, u_c])\}$ is the one we considered in the previous papers. As suggested for the one variable queries, asking for a region is less interesting, but asking for the smallest probability interval may be useful. For example, the query $\{\langle id, [\ell, u]\rangle \mid (id, q_c, t_c, [\ell, u])\}$ asks for all the objects and their smallest probability intervals such that the object is in region q_c at time t_c with that probability interval. Then, $\{\langle t, [\ell, u]\rangle \mid (id_c, q_c, t, [\ell, u])\}$ is interpreted in a similar way. Among the three variable queries probably the most interesting one is $\{\langle id, t, [\ell, u]\rangle \mid (id, q_c, t, [\ell, u])\}$ which essentially asks for all information about a region, that is, what objects were in the region at what times with what probability intervals.

Next we consider queries with connectives. Although it is not necessary to use the same selection queries that we used in the previous papers, we will do so here. Starting with conjunction, consider the query $\{\langle id, t\rangle \mid (id, q_c, t, [\ell_c, u_c]) \wedge (id, q_d, t, [\ell_d, u_d])\}$ asking for all $\langle id, t\rangle$ pairs such that the object was at that time both in region q_c with probability in the interval $[\ell_c, u_c]$ and in the region q_d with probability in the interval $[\ell_d, u_d]$. The result will be the intersection of two answers to two atomic queries. Similarly, the query $\{\langle id, t\rangle \mid (id, q_c, t, [\ell_c, u_c]) \vee (id, q_d, t, [\ell_d, u_d])\}$ gives the union of two atomic queries, and $\{\langle id, t\rangle \mid \neg(id, q_c, t, [\ell_c, u_c])\}$ gives the complement of the query answer without the negation but only if we also switch between optimistic and cautious selection. Finally, we add quantifiers. We consider only two examples. The query $\{\langle id\rangle \mid \exists t(id, q_c, t, [\ell_c, u_c])\}$ gives all objects that were at some time value in the region q_c with the probability in $[\ell_c, u_c]$. The same query with the universal quantifier, namely $\{\langle id\rangle \mid \forall t(id, q_c, t, [\ell_c, u_c])\}$ gives all objects that were in the region q_c with probability in $[\ell_c, u_c]$ for all time values.

7.2 Aggregates and Views

[57] is the definitive work on aggregates in probabilistic databases. A typical example in that paper concerns a day-ahead energy market with probabilities for various

prices. Basic aggregate queries involve finding the mean or standard deviation over prices for specified values of other attributes. In the SPOT framework this does not apply directly as we don't actually deal with numeric values (although we use integers for time values). Even so we may want to count the number of objects in a certain region at a certain time. Here we would compute the expected value based on the probabilities. The research problem is the application of the algorithms developed earlier for probabilistic aggregates in the SPOT framework.

Actually, we could add genuine numeric values to SPOT . Consider that there may be a benefit value given to the presence of an object being in a certain region at a certain time. This might be helpful, say, in a military operation or in the case of a taxi service. In this case we would augment the standard 4 columns by adding a fifth column of numeric values. We could then try to solve the problem of answering typical aggregate queries in an efficient way.

Views have been found very useful in relational databases. In the SPOT framework the most useful views would be obtained by asking a selection query that preserves all the attributes, so that the end result is also a SPOT database. Here the query $\{\langle id_c, q, t, [\ell, u]\rangle\}$ selects all the SPOT atoms for id_c. Thus one user may have a view with information only about a single object. Standard questions about views involve answering queries using views and view maintenance. In this simple example the view given can answer only queries about one object, id_c (it gives the null result for all other objects). View maintenance may involve atoms not in the view in case there are obstacles and speed issues concerning the movement of objects that in previous papers we modeled by using a reachability definition. The challenge here is to check when this needs to be done and find ways to speed up the process.

7.3 Groups of Objects and Time Intervals

In the SPOT framework some of the atomic values represent a set. In particular, a region is a set of points in *Space*; we have dealt mainly with rectangular regions, but only for ease of implementation. Also, for probabilities we allow probability intervals. Now we show how to extend SPOT for objects and time intervals.

In the case of objects we may be interested in a group of objects. For example, a military convoy may consist of multiple vehicles moving together. We can use the notation $(G, q, t, [\ell, u])$ to mean that the group of objects G is in region q at time t with probability in $[\ell, u]$. In such a case, if say $G = \{id_1, id_2, id_3\}$, we could have written this in SPOT as $(id_i, q, t, [\ell, u])$ for $i = 1, 2, 3$, that is, as three SPOT atoms. But there may be another meaning assigned to the atom $(G, q, t, [\ell, u])$: namely, that some object in G is in the region q at time t with probability interval $[\ell, u]$. The latter is the disjunctive interpretation of G, while the former is the conjunctive interpretation. The disjunctive interpretation cannot be expressed in SPOT .

In the case of time it would be natural to extend a single time value to a time interval. Consider the notation $(id, q, [t_1, t_2], [\ell, u])$ to indicate that the object id is in region q at times between t_1 and t_2 with probability in $[\ell, u]$. Again, there is a conjunctive and a disjunctive interpretation. In the conjunctive interpretation the object is in the region with the given probability interval for all time values between t_1 and t_2, while in the disjunctive interpretation it is the case for some time value between t_1 and t_2.

The two cases, objects and times, can be combined to obtain the new generalized atom $(G,q,[t_1,t_2],[\ell,u])$ where there are four possible meanings using the conjunctive and disjunctive interpretations. The research challenge then is to extend the previous SPOT results to such generalized atoms. This involves answering various types of queries and doing database updates in the sense of revision. An interesting question is what data structures will be useful for speeding up processing in these cases.

7.4 *Full* SPOT *Logic*

The SPOT framework allows only atomic formulas, but these formulas have a strong expressive capability. Still, it is reasonable to consider what can be done if we allow connectives and quantifiers in the language. Actually, in such a case it will be useful also to allow both open and half-open probability intervals, so that, for example, $(id,q,t,(\ell,u])$ states that *id* is in region q at time t with probability interval greater than ℓ and at most u.

Starting with conjunction we note that a conjunction of SPOT atoms, $a_1 \wedge \ldots \wedge a_n$ is equivalent to the set $\{a_1,\ldots,a_n\}$. There is a difference however in updating a SPOT database. Recall that in order to retain consistency we introduced several strategies (see Section 5). Suppose that the conflict of a new atom is with a_1 only. Using maximal consistent set revision with $\{a_1,\ldots,a_n\}$ we delete a_1, but if we have the conjunction only, we must delete it, thereby losing the information in the other $a_i, i=2,\ldots,n$. Even if we apply a different strategy, such as minimizing probability, the definitions must be rewritten allowing, for example, in $a_1 \wedge \ldots \wedge a_n$ to change only one probability interval.

Disjunction presents substantial challenges (but then the same goes for going from relational to disjunctive databases). It is not clear that we can write a single linear program where the existence of a solution signifies the consistency of a disjunctive database, such as, if the disjunction $(id_1,q_1,t_1,[\ell_1,u_1]) \vee (id_2,q_2,t_2,[\ell_2,u_2])$ is included. Thus it may not be possible to adapt our approach using linear programming for this case. So it may be appropriate to limit disjunction in certain ways. For example, we may allow only one of the values to change, as in $(id_1,q,t,[\ell,u]) \vee (id_2,q,t,[\ell,u])$ where the identity of the object is known to be id_1 or id_2. Actually, this is the same situation that we considered in the previous subsection where we allowed groups of objects. However, the disjunction $(id,q,t,[\ell_1,u_1]) \vee (id,q,t,[\ell_2,u_2])$ assuming that $u_1 < \ell_2$ is not covered there.

Let's consider negation next. Here is where our extension to half-open intervals becomes useful. Essentially, the negation of an atom in SPOT can be expressed in the form of a disjunction, as follows: $\neg(id,q,t,[\ell,u]) \equiv (id,q,t,[0,\ell)) \vee (id,q,t,(u,1])$, with the appropriate modification in case $\ell = 0$ or $u = 1$. In that sense negation can be eliminated in full SPOT logic.

Finally, we take a look at quantifiers. In case *Id*, *Space* and *Time* are finite and we do not quantify over probabilities, we can rewrite quantified statements using conjunction and disjunction. For example, if $T = \{0, \ldots, N\}$, the statement $\exists t(id_c, q_c, t, [\ell_c, u_c])$, expressing that the object id_c was in region q_c at some time value with probability in the interval $[\ell_c, u_c]$ can be written without the existential quantifier as the disjunction $(id_c, q_c, 0, [\ell_c, u_c]) \vee \ldots \vee (id_c, q_c, N, [\ell_c, u_c])$. Similarly, changing the existential quantifier to a universal quantifier results in the disjunction changed to a conjunction. However, if we choose T to be infinite, then the quantifier cannot be replaced.

7.5 Adding Spatial and Temporal Concepts

Space and time are essential components in the SPOT framework. However, although SPOT is based on logical formalism, it does not take into account any concepts from the logics that have been developed for space and time. This subsection discusses some ideas that that could be used to augment the SPOT framework with spatial and temporal logic concepts.

We refer to the survey paper [32] for logical formalisms concerning spatial concepts. There are formal logical systems for spatial relations between regions such as $part\_of$ and *overlap* as well as spatial properties of regions such as *open*, *closed*, and *connected*, as well as operators such as *closure* and *boundary*. Thus we propose that this extension contain the usual SPOT atoms as well as atomic formulas that represent spatial concepts, and a proper mixture of the two. We assume here that regions are not restricted to rectangles in *Space*; in fact *Space* itself can be any set of points.

Allowing O to represent the overlap relation, the following three atoms: $(id, q, t, [\ell, u])$; $(id, r, t, [\ell, u])$; $O(q, r)$ clearly give more information than just the first two. We would also like to allow operators inside SPOT atoms; for example $(id, B(q), t, [\ell, u])$, where B stands for boundary, indicates the probability interval for the object *id* to be on the boundary of region q at time t. For connectedness, in the case of moving objects we may wish to define several types, such as a region is *car-connected* if one can travel by car from any point to any other point in the region versus *plane-connected* where travel is by plane. The research here involves the best spatial extensions and algorithms to deal with them.

Important issues about temporal logic are well summarized in [26]. As explained there, temporal logics have temporal operators such as F: "It will at some time be the case that ... " and G: "It will always be the case that ... ". Using such temporal operators to make general statements we may allow SPOT statements without the time component to write $G(id, q, [.8, 1])$ to mean that the object *id* will always be in the region q with probability at least .8. If time were finite, say $T = \{0, 1, \ldots, N\}$ and the present time *now* is known, then we could have written the $N -$ *now* statements $(id, q, n, [.8, 1])$ for $n = now + 1, \ldots, N$ with the same meaning. However, this cannot be done with our assumption that T is the set of integers. Also, even with finite time we cannot express $F(id,q,[.8,1])$. So, in general, temporal operators expand the expressibility of the SPOT framework.

In the previous subsection we considered the addition of quantifiers to SPOT . Using quantifiers and inequality predicates we can express the temporal operators. For example, $F(id,q,[.8,1])$ would be written as $\exists t(t > now \wedge (id,q,t,[.8,1])$; G is the same but with the universal quantifier. However, quantifiers can cause complications and if our interest is only in expressing various temporal concepts, it may be better to use temporal operators. The important issue is the development of data structures to process such concepts as efficiently as possible. Finally, we mention the combination of spatial and temporal concepts that might be added to the SPOT framework for widening its expressibility.

References

1. Agarwal, P.K., Arge, L., Erickson, J.: Indexing moving points. Journal of Computer and System Sciences 66(1), 207–243 (2003)
2. Aziz, A., Sanwal, K., Singhal, V., Brayton, R.K.: Verifying continuous time markov chains. In: Alur, R., Henzinger, T.A. (eds.) CAV 1996. LNCS, vol. 1102, pp. 269–276. Springer, Heidelberg (1996)
3. Barbará, D., Garcia-Molina, H., Porter, D.: The management of probabilistic data. IEEE TKDE 4(5), 487–502 (1992)
4. Benjelloun, O., Sarma, A.D., Halevy, A.Y., Widom, J.: Uldbs: Databases with uncertainty and lineage. In: VLDB, pp. 953–964 (2006)
5. Bennett, B.: Modal logics for qualitative spatial reasoning. Journal of the Interest Group on Pure and Applied Logic 4, 23–45 (1996)
6. Brusoni, V., Console, L., Terenziani, P., Pernici, B.: Extending temporal relational databases to deal with imprecise and qualitative temporal information. In: Clifford, S., Tuzhilin, A. (eds.) Intl. Workshop on Recent Advances in Temporal Databases, pp. 3–22. Springer (1995)
7. Cao, H., Wolfson, O., Trajcevski, G.: Spatio-temporal data reduction with deterministic error bounds. VLDB Journal 15, 211–228 (2006)
8. Cavallo, R., Pittarelli, M.: The theory of probabilistic databases. In: VLDB, pp. 71–81 (1987)
9. Chen, Y.F., Qin, X.L., Liu, L.: Uncertain distance-based range queries over uncertain moving objects. J. Comput. Sci. Technol. 25(5), 982–998 (2010)
10. Chung, B.S.E., Lee, W.C., Chen, A.L.P.: Processing probabilistic spatio-temporal range queries over moving objects with uncertainty. In: EDBT, pp. 60–71 (2009)
11. Cohn, A.G., Hazarika, S.M.: Qualitative spatial representation and reasoning: an overview. Fundam. Inf. 46(1-2), 1–29 (2001)
12. Dai, X., Yiu, M.L., Mamoulis, N., Tao, Y., Vaitis, M.: Probabilistic spatial queries on existentially uncertain data. In: Medeiros, C.B., Egenhofer, M., Bertino, E. (eds.) SSTD 2005. LNCS, vol. 3633, pp. 400–417. Springer, Heidelberg (2005)
13. Dalvi, N.N., Suciu, D.: Efficient query evaluation on probabilistic databases. VLDB J 16(4), 523–544 (2007)
14. Dekhtyar, A., Dekhtyar, M.I.: Possible worlds semantics for probabilistic logic programs. In: Demoen, B., Lifschitz, V. (eds.) ICLP 2004. LNCS, vol. 3132, pp. 137–148. Springer, Heidelberg (2004)

15. Dey, D., Sarkar, S.: A probabilistic relational model and algebra. ACM Trans. Database Syst. 21(3), 339–369 (1996)
16. Dubois, D., Prade, H.: Processing fuzzy temporal knowledge. IEEE Transactions on Systems, Man and Cybernetics 19(4), 729–744 (1989)
17. Dutta, S.: Generalized events in temporal databases. In: Proc. 5th Intl. Conf. on Data Engineering, pp. 118–126 (1989)
18. Eiter, T., Lukasiewicz, T., Walter, M.: A data model and algebra for probabilistic complex values. Ann. Math. Artif. Intell. 33(2-4), 205–252 (2001)
19. Fagin, R., Halpern, J.Y., Megiddo, N.: A logic for reasoning about probabilities. Inf. Comput. 87(1/2), 78–128 (1990)
20. Fusun Yaman, D.N., Subrahmanian, V.: The logic of motion. In: Proc. 9th International Conference on the Principles of Knowledge Representation and Reasoning (KR 2004), pp. 85–94 (2004)
21. Fusun Yaman, D.N., Subrahmanian, V.: Going far, logically. In: Proc. IJCAI 2005, pp. 615–620 (2005)
22. Fusun Yaman, D.N., Subrahmanian, V.: A motion closed world assumption. In: Proc. IJCAI 2005, pp. 621–626 (2005)
23. Cohn, A.G., Magee, D.R., Galata, A., Hogg, D.C., Hazarika, S.M.: Towards an architecture for cognitive vision using qualitative spatio-temporal representations and abduction. In: Freksa, C., Brauer, W., Habel, C., Wender, K.F. (eds.) Spatial Cognition III. LNCS (LNAI), vol. 2685, pp. 232–248. Springer, Heidelberg (2003)
24. Gabelaia, D., Kontchakov, R., Kurucz, A., Wolter, F., Zakharyaschev, M.: On the computational complexity of spatio-temporal logics. In: FLAIRS Conference, pp. 460–464 (2003)
25. Gadia, S., Nair, S., Poon, Y.: Incomplete infromation in relational temporal databases. In: Proc. Intl. Conf. on Very Large Databases (1992)
26. Galton, A.: Temporal logic. In: Zalta, E.N. (ed.) The Stanford Encyclopedia of Philosophy, fall 2008 edn. (2008)
27. Grant, J., Parisi, F., Parker, A., Subrahmanian, V.S.: An agm-style belief revision mechanism for probabilistic spatio-temporal logics. Artif. Intell. 174(1), 72–104 (2010)
28. Güntzer, U., Kiessling, W., Thöne, H.: New direction for uncertainty reasoning in deductive databases. In: Proceedings of the 1991 ACM SIGMOD International Conference on Management of Data, SIGMOD 1991, pp. 178–187. ACM, New York (1991), http://doi.acm.org/10.1145/115790.115815
29. Hadjieleftheriou, M., Kollios, G., Tsotras, V.J., Gunopulos, D.: Efficient indexing of spatiotemporal objects. In: Jensen, C.S., Jeffery, K., Pokorný, J., Šaltenis, S., Bertino, E., Böhm, K., Jarke, M. (eds.) EDBT 2002. LNCS, vol. 2287, pp. 251–268. Springer, Heidelberg (2002)
30. Hansson, H., Jonsson, B.: Logic for reasoning about time and reliability. Formal Asp. Comput. 6(5), 512–535 (1994)
31. Jampani, R., Xu, F., Wu, M., Perez, L.L., Jermaine, C., Haas, P.J.: MCDB: a monte carlo approach to managing uncertain data. In: Proceedings of the 2008 ACM SIGMOD International Conference on Management of Data, pp. 687–700. ACM, New York (2008)
32. Jeansoulin, R., Papini, O., Prade, H., Schockaert, S.: Introduction: uncertainty issues in spatial information. In: Jeansoulin, R., Papini, O., Prade, H., Schockaert, S. (eds.) Methods for Handling Imperfect Spatial Information. STUDFUZZ, vol. 256, pp. 1–14. Springer, Heidelberg (2010)
33. Kifer, M., Li, A.: On the semantics of rule-based expert systems with uncertainty. In: ICDT, pp. 102–117 (1988)

34. Koch, C., Olteanu, D.: Conditioning probabilistic databases. Proceedings of the VLDB Endowment archive 1(1), 313–325 (2008)
35. Kollios, G., Gunopulos, D., Tsotras, V.J.: On indexing mobile objects. In: Proceedings of the eighteenth ACM SIGMOD-SIGACT-SIGART Symposium on Principles of Database Systems, pp. 261–272. ACM, New York (1999)
36. Koubarakis, M.: Database models for infinite and indefinite temporal information. Information Systems 19(2), 141–173 (1994)
37. Kwiatkowska, M., Norman, G., Parker, D.: PRISM: Probabilistic symbolic model checker. In: Field, T., Harrison, P.G., Bradley, J., Harder, U. (eds.) TOOLS 2002. LNCS, vol. 2324, pp. 200–204. Springer, Heidelberg (2002)
38. Lakshmanan, L.V., Sadri, F.: Modeling uncertainty in deductive databases. In: Karagiannis, D. (ed.) DEXA 1994. LNCS, vol. 856, pp. 724–733. Springer, Heidelberg (1994)
39. Lakshmanan, L.V.S., Shiri, N.: A parametric approach to deductive databases with uncertainty. IEEE Trans. on Knowl. and Data Eng. 13(4), 554–570 (2001)
40. Lian, X., Chen, L.: Probabilistic group nearest neighbor queries in uncertain databases. IEEE Transactions on Knowledge and Data Engineering 20(6), 809–824 (2008), `http://doi.ieeecomputersociety.org/10.1109/TKDE.2008.41`
41. Lukasiewicz, T.: Probabilistic logic programming. In: ECAI, pp. 388–392 (1998)
42. Lukasiewicz, T., Kern-Isberner, G.: Probabilistic logic programming under maximum entropy. In: Hunter, A., Parsons, S. (eds.) ECSQARU 1999. LNCS (LNAI), vol. 1638, pp. 279–292. Springer, Heidelberg (1999)
43. Merz, S., Wirsing, M., Zappe, J.: A spatio-temporal logic for the specification and refinement of mobile systems. In: Pezzé, M. (ed.) FASE 2003. LNCS, vol. 2621, pp. 87–101. Springer, Heidelberg (2003)
44. Muller, P.: Space-Time as a Primitive for Space and Motion. In: FOIS, pp. 63–76. IOS Press, Amsterdam (1998), `http://www.irit.fr/~Philippe.Muller`
45. Ng, R.T., Subrahmanian, V.S.: Probabilistic logic programming. Information and Computation 101(2), 150–201 (1992), `citeseer.csail.mit.edu/ng92probabilistic.html`
46. Ni, J., Ravishankar, C.V., Bhanu, B.: Probabilistic spatial database operations. In: Hadzilacos, T., Manolopoulos, Y., Roddick, J., Theodoridis, Y. (eds.) SSTD 2003. LNCS, vol. 2750, pp. 140–159. Springer, Heidelberg (2003)
47. Parisi, F., Parker, A., Grant, J., Subrahmanian, V.S.: Scaling cautious selection in spatial probabilistic temporal databases. In: Jeansoulin, R., Papini, O., Prade, H., Schockaert, S. (eds.) Methods for Handling Imperfect Spatial Information. STUDFUZZ, vol. 256, pp. 307–340. Springer, Heidelberg (2010)
48. Parker, A., Infantes, G., Grant, J., Subrahmanian, V.: An agm-based belief revision mechanism for probabilistic spatio-temporal logics. In: AAAI (2008)
49. Parker, A., Infantes, G., Grant, J., Subrahmanian, V.S.: Spot databases: Efficient consistency checking and optimistic selection in probabilistic spatial databases. IEEE TKDE 21(1), 92–107 (2009)
50. Parker, A., Subrahmanian, V.S., Grant, J.: A logical formulation of probabilistic spatial databases. IEEE TKDE, 1541–1556 (2007)
51. Parker, A., Yaman, F., Nau, D., Subrahmanian, V.: Probabilistic go theories. In: IJCAI, pp. 501–506 (2007)
52. Pelanis, M., Saltenis, S., Jensen, C.S.: Indexing the past, present, and anticipated future positions of moving objects. ACM Trans. Database Syst. 31(1), 255–298 (2006)
53. Pfoser, D., Jensen, C.S., Theodoridis, Y.: Novel approaches to the indexing of moving object trajectories. In: Proceedings of VLDB (2000)
54. Pittarelli, M.: An algebra for probabilistic databases. IEEE TKDE 6(2), 293–303 (1994)

55. Rajagopalan, R., Kuipers, B.: Qualitative spatial reasoning about objects in motion: Application to physics problem solving. In: IJCAI 1994, San Antonio, TX, pp. 238–245 (1994)
56. Randell, D.A., Cui, Z., Cohn, A.G.: A spatial logic based on regions and connection. In: International Conference on Knowledge Representation and Reasoning, KR 1992, pp. 165–176. Morgan Kaufmann (1992)
57. Ross, R., Subrahmanian, V.S., Grant, J.: Aggregate operators in probabilistic databases. Journal of the ACM 52(1), 54–101 (2005)
58. Shanahan, M.: Default reasoning about spatial occupancy. Artif. Intell. 74(1), 147–163 (1995), http://dx.doi.org/10.1016/0004-3702(94)00071-8
59. Snodgrass, R.: The temporal query language tquel. In: Proceedings of the 3rd ACM SIGACT-SIGMOD Symposium on Principles of Database Systems, PODS 1984, pp. 204–213. ACM, New York (1984), http://doi.acm.org/10.1145/588011.588041
60. Tao, Y., Cheng, R., Xiao, X., Ngai, W.K., Kao, B., Prabhakar, S.: Indexing multi-dimensional uncertain data with arbitrary probability density functions. In: VLDB, pp. 922–933 (2005)
61. Tao, Y., Papadias, D., Sun, J.: The TPR*-tree: an optimized spatio-temporal access method for predictive queries. In: Proceedings of the 29th International Conference on Very Large Data Bases, vol. 29, pp. 790–801. VLDB Endowment (2003)
62. Wolter, F., Zakharyaschev, M.: Spatial reasoning in rcc-8 with boolean region terms. In: Principles of Knowledge Representation and Reasoning, ECAI 2000, pp. 244–248. IOS Press, Berlin (2000)
63. Wolter, F., Zakharyaschev, M.: Spatio-temporal representation and reasoning based on RCC-8. In: Cohn, A.G., Giunchiglia, F., Selman, B. (eds.) Principles of Knowledge Representation and Reasoning, KR 2000, pp. 3–14. Morgan Kaufmann, San Francisco (2000), citeseer.ist.psu.edu/wolter00spatiotemporal.html
64. Yang, B., Lu, H., Jensen, C.S.: Probabilistic threshold k nearest neighbor queries over moving objects in symbolic indoor space. In: Manolescu, I., Spaccapietra, S., Teubner, J., Kitsuregawa, M., Léger, A., Naumann, F., Ailamaki, A., Özcan, F. (eds.) EDBT. ACM International Conference Proceeding Series, vol. 426, pp. 335–346. ACM (2010)
65. Zhang, M., Chen, S., Jensen, C.S., Ooi, B.C., Zhang, Z.: Effectively indexing uncertain moving objects for predictive queries. PVLDB 2(1), 1198–1209 (2009)
66. Zheng, K., Trajcevski, G., Zhou, X., Scheuermann, P.: Probabilistic range queries for uncertain trajectories on road networks. In: Ailamaki, A., Amer-Yahia, S., Patel, J.M., Risch, T., Senellart, P., Stoyanovich, J. (eds.) EDBT, pp. 283–294. ACM (2011)

A Probabilistic Object-Oriented Database Model with Fuzzy Measures

Li Yan and Z. Ma

Abstract. This chapter concentrates on modeling probabilistic evens with fuzzy measures in the object-oriented databases. Instead of crisp probability measures or interval probability measures of objects and classes, fuzzy sets are applied to represent imprecise probability measures. A probabilistic object-oriented database model with fuzzy measures is introduced, which incorporates fuzzy probability degrees to handle imprecision and uncertainty. Based on the proposed probabilistic object-oriented database model, several major semantic relationships of probabilistic objects and probabilistic classes with fuzzy measures are investigated in the chapter.

1 Introduction

Databases are used to manage an enormous wealth of data from various real-world applications. As a result, one of the major areas of database research has been the continuous effort to enrich existing database models with a more extensive collection of semantic concepts in order to satisfy the requirements in the real-world applications. Database models have developed from hierarchical and network database models to the relational database model. As computer technology moves into non-traditional applications (e.g. knowledge-based systems), many software engineers feel the limitations of relational databases in data- and knowledge-intensive applications. So some non-traditional data models (e.g., the object-oriented database model and the object-relational database model) have been proposed for databases.

Li Yan
School of Software, Northeastern University, Shenyang, 110819, China

Z. Ma
School of Information Science & Engineering, Northeastern University, Shenyang, 110819, China
e-mail: mazongmin@ise.neu.edu.cn

Z. Ma & L. Yan (Eds.): *Advances in Probabilistic Databases*, STUDFUZZ 304, pp. 23–38.
DOI: 10.1007/978-3-642-37509-5_2

While the object-oriented database model provides powerful object-oriented modeling capabilities, it suffers from some inadequacy of necessary semantics. In real-world applications, information is often imperfect, and human knowledge and natural language have a big deal of imprecision and vagueness. Traditional database models assume that the models are a correct reflection of the world and further assume that the stored data is known, accurate and complete. It is rarely the case in real life that all or most of these assumptions are met. One of some inadequacy of necessary semantics that traditional database models often suffer from can hereby be generalized as the inability to handle imprecise and uncertain information. For this reason, imprecise and uncertain data have been introduced into databases for imperfect information processing [10, 15, 16, 21].

To deal with uncertainty and imprecision, two major foundations have been developed, which are probability theory and fuzzy set theory, respectively. Fuzzy set theory and probability theory have been applied to extend various database models and this has resulted in numerous contributions, mainly with respect to the popular relational model. But the fuzzy relational database model [3, 7, 13, 14, 17, 23, 24, 25, 27] or probabilistic relational database model [6, 8, 12, 22, 32] does not satisfy the requirement of modeling complex objects with inherent imprecision and uncertainty. The object-oriented database model can represent complex object structures without fragmenting the aggregate data and can also depict complex relationships among attributes. Recently some efforts have concentrated on the fuzzy object-oriented databases [18, 26] and probabilistic object-oriented databases [9, 11, 20]. As a result, some related notions (e.g., class, superclass/subclass, inheritance and algebraic operations) have been investigated. While the fuzzy object-oriented databases and probabilistic object-oriented databases have been developed separately to deal with subjective uncertainty and objective uncertainty, respectively, each of these two object-oriented database models actually suffers from the inability to simultaneously handle fuzzy and probabilistic information. In a real-world application of weather forecast, for example, it is common to say "*the probability it will be mainly sunny tomorrow is 0.90*". But "*the probability it will be mainly sunny tomorrow is very high*" may be easily understood by human beings. Here we have to face a probabilistic event which probability is a fuzzy one instead of a crisp one. At this point, the interaction of fuzzy set theory and probability theory is needed and they are complementary to each other [1].

Actually imprecise probability measures have been investigated in literature. In [12], imprecise probability measures represented by interval probabilities were introduced into the probabilistic relational databases, where two additional attributes are needed to represent the minimum and maximum probabilities of tuples. Interval probabilities were also applied to extend the object-oriented databases in [9]. Also there are some efforts to combine fuzzy set theory and probability theory together. In [5], a deductive probabilistic and fuzzy object-oriented database language was proposed, where a class property can contain fuzzy set values, and uncertain class membership and property applicability are measured by lower and upper bounds on probability. Similarly an object base model was developed in [4], which incorporates both fuzzy set values and probability degrees represented by interval probabilities to cope with imprecision and uncertainty. It is shown that in order to model imprecision in probabilistic evens in databases, fuzzy sets have been applied to imprecise attribute values. As to imprecise probability measures, only interval values have been considered and investigated. For descriptions of imprecise information, we know that a fuzzy value represented by a fuzzy set contains more informative semantics than an interval value. It is a natural way to use fuzzy values to represent probability measures which are imperfectly known.

In this chapter, we concentrate on modeling probabilistic evens with fuzzy measures in the object-oriented databases. Instead of crisp probability measures of objects in [11] and interval probability measures of objects in [4, 9], we use fuzzy sets to represent imprecise probability measures. A probabilistic object-oriented database model with fuzzy measures is hence introduced, which incorporates fuzzy probability degrees to handle imprecision and uncertainty. Based on the proposed probabilistic object-oriented database model, we discuss several major semantic relationships between objects, objects and classes as well as classes.

The remainder of this chapter is organized as follows. Section 2 provides basic knowledge concerning object-oriented database model and fuzzy sets. Two major concepts are introduced to the probabilistic object-oriented databases in Section 3, which are probabilistic objects with fuzzy measures and probabilistic classes with fuzzy measures, respectively. Section 4 investigates semantic relationship between objects, objects and classes as well as classes in the probabilistic object-oriented database model with fuzzy probability measures. A formal probabilistic object-oriented database model with fuzzy measures is developed in Section 5. Section 6 concludes this chapter.

2 Basic Knowledge

This section provides basic knowledge about object-oriented databases and fuzzy set theory.

2.1 Object-Oriented Databases

An *object-oriented database model* (OODM) is based on the notions of *object*, *class*, *attribute* and *method*. An object is an entity of the real world and is characterized by a unique object identity (*OID*). An object has a state (attributes) and a behavior (methods) associated with it. Objects sharing a common structure and behavior are grouped into the same class. Each object can be an instance of a class or several classes. Classes can be organized into hierarchies of the kind superclass-subclass. A new class, called subclass, is produced from another class, called superclass by means of inheriting all attributes and methods of the superclass, overriding some attributes and methods of the superclass, and defining some new attributes and methods. Any object belonging to the subclass must belong to the superclass.

An object-oriented database model is a finite set of *class declarations*, and each class declaration describes a set of objects sharing a collection of features. Formally an object-oriented database model is a tuple $D = (C, A, R, O, M, S)$, in which C is a finite set of classes, A is a finite set of attributes, R is a finite set of relationships, O is a finite set of objects, M is a finite set of methods, and S is a finite set of constraints. This paper focuses on the classes, attributes, relationships, and objects, yielding a simple object-oriented database model: $D = (C, A, R, O)$, where $C = \{c_1, c_2, \ldots, c_k\}$, $A = \{a_1, a_2, \ldots, a_l\}$, $R = \{r_1, r_2, \ldots, r_m\}$, and $O = \{o_1, o_2, \ldots, o_n\}$. Then, we have:

- $R \subseteq C \times C$ is a binary relation that represents subclass-superclass relationship.
- For $c_i \in C$ $(1 \le i \le k)$, A (c_i) represents a set of attributes of c_i. Clearly A (c_i) $\subseteq \{a_1, a_2, \ldots, a_l\}$, i.e., A $(c_i) \subseteq A$. For $a_j \in A$ $(1 \le j \le l)$, a_j (c_i) denotes the attribute a_j of c_i. In the context of the given c_i, a_j is used instead of a_j (c_i). In addition, for $a_j \in A$ $(1 \le j \le l)$, there exits a domain, denoted by *dom* (a_j).
- For $c_i \in C$ $(1 \le i \le k)$, O (c_i) means a set of objects that c_i contains. Here, O $(c_i) \subseteq \{o_1, o_2, \ldots, o_n\}$, i.e., O $(c_i) \subseteq O$. For $o_p \in O$ $(1 \le p \le n)$ and $a_j \in A$ $(1 \le j \le l)$, o_p (c_i) denotes the object o_p of c_i, and o_p $(a_j$ $(c_i))$ denotes the value of object o_p on attribute a_j. In the context of the given c_i, o_p is used instead of o_p (c_i) and o_p (a_j) is used instead of o_p $(a_j$ $(c_i))$. It is clear that o_p $(a_j) \in$ Dom (a_j).

2.2 Fuzzy Sets

The concept of fuzzy sets was originally introduced by Zadeh [28]. Let U be a universe of discourse and F be a fuzzy set in U. A membership function μ_F: $U \rightarrow [0, 1]$ is defined for F, where $\mu_F(u)$ for each $u \in U$ denotes the membership degree of u in the fuzzy set F. Thus, the fuzzy set F is described as follows:

$$F = \{(u_1, \mu_F(u_1)), (u_2, \mu_F(u_2)), \ldots, (u_n, \mu_F(u_n))\}$$

In fuzzy set F, each element may or may not belong to F and has a membership degree to F that needs to be explicitly indicated. So an element in F (say u_i) is associated with its membership degree (say $\mu_F(u_i)$) and these occur together in the form of $\mu_F(u_i)/u_i$. When the membership degrees indicate that all elements in F belong to F with membership degrees of exactly 1, fuzzy set F reduces to the conventional set.

When the membership degree $\mu_F(u)$ above is explained as a measure of the possibility that a variable X has the value u, where X takes on values in U, a fuzzy value is described by the possibility distribution π_X [30].

$$\pi_X = \{(u_1, \pi_X(u_1)), (u_2, \pi_X(u_2)), ..., (u_n, \pi_X(u_n))\}$$

Here, $\pi_X(u_i)$ ($u_i \in U$ and $0 \leq i \leq n$) denotes the possibility that u_i is true. Let π_X be the representation of the possibility distribution for the fuzzy value of a variable X. This means that the value of X is fuzzy, and X may take on one of the possible values u_1, u_2, ..., and u_n, with each possible value (say u_i) associated with a possibility degree (say $\pi_X(u_i)$).

The set of the elements in π_X which possibilities are non-zero is called the support of π_X, denoted by

$$\text{supp}(\pi_X) = \{u | u \in U \text{ and } \pi_X(u) > 0\}.$$

The extension principle introduced by Zadeh [29] is regarded as one of the most basic ideas of fuzzy set theory. By providing a general method, the extension principle has been extensively employed to extend nonfuzzy mathematical concepts. The idea is to induce a fuzzy set from a number of given fuzzy sets using a mapping. Zadeh's extension principle can also be sometimes referred as maximum-minimum principle. Let π_A and π_B be two fuzzy data based on possibility distribution on the universe of discourse $U = \{u_1, u_2, \ldots, u_n\}$, and π_A and π_B are respectively represented by

$$\pi_A = \{\pi_A(p_i)/p_i | p_i \in U \wedge 1 \leq i \leq n\} \text{ and } \pi_B = \{\pi_B(q_j)/q_j | q_j \in U \wedge 1 \leq j \leq n\}.$$

Following the Zadeh's extension principle, the operation with an infix operator "θ" on π_A and π_B can be defined as follows.

$$\pi_A \,\theta\, \pi_B = \{\pi_A(p_i)/p_i | u_i \in U \wedge 1 \leq i \leq n\} \,\theta\, \{\pi_B(q_j)/q_j | q_j \in U \wedge 1 \leq j \leq n\} = \{\max(\min(\pi_A(p_i), \pi_B(q_j))/p_i \,\theta\, q_j) | p_i, q_j \in U \wedge 1 \leq i, j \leq n\}$$

3 Probabilistic Objects and Probabilistic Classes with Fuzzy Measures

The notions of objects and classes are the core of the object-oriented databases. In order to model and handle fuzzy and probabilistic information in the object-oriented databases, objects and classes must be extended. This section introduces probabilistic objects with fuzzy measures and probabilistic classes with fuzzy measures.

3.1 Probabilistic Objects with Fuzzy Measures

Probabilistic information has been introduced into relational databases, in which attribute values may be probability distributions [2] or tuples may be associated with probability measures [8, 12, 22, 32]. In [8], a crisp probability measure associated with a tuple is actually the joint probability of the given realizations of all the attributes taken together. For this purpose, an additional attribute *pS* is introduced to the probabilistic relational scheme, where Dom (*pS*) = [0, 1]. In order to represent imprecise probability measures of tuples, i.e., interval probability measures, in [12], two additional attributes *LB* and *UB* are introduced into the probabilistic relational scheme. These two attributes are used to represent the lower boundary and upper boundary of probability measures of tuples. It is clear that Dom (*LB*) = [0, 1] and Dom (*UB*) = [0, 1].

Viewed from the object-oriented paradigm, a complex event in the real world traditionally corresponds to an object which describes the status of the event. But a complex and stochastic event in the real world may have multiple statuses and each status should be described by an object. So a complex and stochastic event in the real world may be described by a group of objects in the context of the object-oriented paradigm, in which each object is associated with a probability measure, indicating the probability that this stochastic event in the real world has the status described by the object. The objects associated with a stochastic event have the same object identifier and represent the complete joint distribution of its attribute values, called *a virtual object*. There is a probabilistic constraint on the objects with the same object identifier (i.e., virtual object): the sum of probability measures of all objects with the same object identifier is less than or equal to one. Formally, let *e* be a complex and stochastic event, which corresponds to a virtual object o_v and is described by a group of objects $\{o_1, o_2, \ldots, o_k\}$. Here $o_1, o_2, \ldots, o_k$ have the same object identifier. Let the probability measure of object o_i $(1 \le i \le k)$ be $o_i\ (p_e)$. Then we have $\sum_{i = 1, 2, \ldots, k} o_i\ (p_e) \le 1$.

The probability measure associated with an object is a crisp one expressed by a precise value when it is definitely known. Also it is possible that the probability measure associated with an object is fuzzily known. At this point, the probability measure associated with the object is a fuzzy probability measure and is expressed by a possibility distribution. Then the above-mentioned probabilistic constraint that the sum of probability measures of all objects with the same object identifier is less than or equal to one should be adjusted. The new constraint on these objects with fuzzy probability measures, called *fuzzy probabilistic constraint*, is that, for all objects with the same identifier, the sum of their maximum joint probabilities must be less than or equal to one. Here, the maximum probability of object o_i is represented by the maximum of the support of $o_i\ (p_e)$.

Formally let $o_1, o_2, \ldots, o_k$ be the objects with the same identifier, and let $o_i\ (p_e) = \{(p_{i1}, \pi_{oi}\ (p_{i1})), (p_{i2}, \pi_{oi}\ (p_{i2})), \ldots, (p_{im}, \pi_{oi}\ (p_{im}))\}$ $(1 \le i \le k)$. Then the fuzzy probabilistic constraint on $o_1, o_2, \ldots, o_k$ is represented as follows.

$$\max (\mathrm{supp}\ (o_1\ (p_e))) + \max (\mathrm{supp}\ (o_2\ (p_e))) + \ldots + \max (\mathrm{supp}\ (o_k\ (p_e))) \le 1.$$

3.2 Probabilistic Classes with Fuzzy Measures

The objects which have the same properties can be gathered into classes. Theoretically, a class can be considered from two different viewpoints:

(1) an extensional class, where the class is defined by the list of its object instances, and
(2) an intensional class, where the class is defined by a set of attributes and their admissible values.

In addition, by means of inheritance mechanism in the object-oriented databases, a subclass defined from its superclass can be seen as a special case of (b) above.

A class is a probabilistic one with a fuzzy measure because of the following several reasons. First, some virtual objects having similar properties are the probabilistic ones with fuzzy measures, and a class defined by these virtual objects may be a probabilistic one with fuzzy measures. Second, when a class is intensionally defined, there is an attribute which domain may be fuzzy and probabilistic, and as a result a probabilistic class with fuzzy measure is formed. Third, the subclass produced by a probabilistic class with fuzzy measure by means of specialization and the superclass produced by some classes (in which there is at least one class which is a probabilistic one with fuzzy measure) by means of generalization are also fuzzy.

A probabilistic class with fuzzy measure contains an additional attribute, which values are applied to indicate the possibilities that the corresponding objects belong to the given class. Also the possibilities may be imprecise values instead of crisp values in [0, 1] and are represented by fuzzy sets. This attribute of the class is called *fuzzy probabilistic attribute*.

The major difference between the probabilistic classes and crisp classes is that the boundaries of the probabilistic classes are uncertain. The uncertainty in the class boundaries is caused by the uncertainty of the values in the attribute domain. In the probabilistic object-oriented databases, an object belongs to a class with a probability represented by a fuzzy set. A class is a subclass of another class with a probability represented by a fuzzy set also.

4 Semantic Relationships of Probabilistic Objects and Classes with Fuzzy Measures

In the object-oriented database model, there are several crucial semantic relationships in objects and classes, including object-object relationship, object-class relationship and class-class relationship. In the probabilistic object-oriented databases with fuzzy measures, the evaluations of probabilistic object-object relationships, probabilistic object-class relationships and probabilistic inheritance hierarchies are the cores of imprecise and uncertain information modeling. In the following, we investigate the semantic relationship of object equivalence, the object-class relationship and subclass-superclass relationship in the context of the probabilistic object-oriented databases with fuzzy measures.

4.1 Probabilistic Object Equivalence

Formally let c be a class of the probabilistic object-oriented database model with fuzzy measures. Let c_1 contain attributes $\{a_1, a_2, \ldots, a_k, fpM\}$, in which fpM is the fuzzy probabilistic attribute of c. For two objects o_1 and o_2 of c, o_1 and o_2 are equivalent each other if and only if $o_1(a_i) = o_2(a_i)$ $(1 \le i \le k)$. It is clear that $a_i \ne fpM$. Two equivalent objects o_1 and o_2 is denoted by $o_1 \approx o_2$.

A virtual object corresponds to a group of objects which have the same object identifier. Suppose that o_1 and o_2 belong to a virtual object. Then o_1 and o_2 are not equivalent objects because at least there exits an attribute (say a_j, where $1 \le j \le k$) and $o_1(a_j) \ne o_2(a_j)$.

Equivalent objects of a class must be coalesced in the probabilistic object-oriented database model with fuzzy measures. As a result, a new object is formed, which probability with fuzzy measure is generated by coalescing the probabilities with fuzzy measures of these equivalent objects. Since the probabilities with fuzzy measures of the equivalent objects contain fuzzy as well as probabilistic information, the coalescence of the probabilities with fuzzy measures of the equivalent objects is calculated using the extension principle [29] according to the fuzzy probabilistic constraint.

For two equivalent objects o_1 and o_2 of class c_1, assume that $o_1(fpM)$ and $o_2(fpM)$ are probability values represented by fuzzy sets. Let $o_1(fpM) = \{(f_1, \mu_1(f_1)), (f_2, \mu_1(f_2)), \ldots, (f_m, \mu_1(f_m))\}$ and $o_2(fpM) = \{(g_1, \mu_2(g_1)), (g_2, \mu_2(g_2)), \ldots, (g_n, \mu_2(g_n))\}$. We define four coalescence operations, which can be used to define projection operation, difference operation, join operation and union operation on probabilistic objects with fuzzy measures, respectively.

- The coalescence-Plus operation. Coalescence-Plus on two equivalent objects o_1 and o_2, denoted $\oplus$, is defined as $o = o_1 \oplus o_2$, in which $o \approx o_1 \approx o_2$ and

  $$o(fpM) = \{\max(\min_{i,j}(1, f_i + g_j), \min_{i,j}(\mu_1(f_i), \mu_2(g_j))) | 1 \le i \le \mathrm{m}, 1 \le j \le \mathrm{n}\}.$$

- The coalescence-Minus operation. The coalescence-Minus on two equivalent objects o_1 and o_2, denoted by Θ, is defined as $o = o_1 \Theta o_2$, in which $o \approx o_1 \approx o_2$ and

  $$o(fpM) = \{\max(\max_{i,j}(f_i - g_j, 0), \min_{i,j}(\mu_1(f_i), \mu_2(g_j))) | 1 \le i \le \mathrm{m}, 1 \le j \le \mathrm{n}\}.$$

- The coalescence-Times operation. The coalescence-Times on two equivalent objects o_1 and o_2, denoted by $\otimes$, is defined as $o = o_1 \otimes o_2$, in which $o \approx o_1 \approx o_2$ and

  $$o(fpM) = \{\max((f_i \times g_j), \min_{i,j}(\mu_1(f_i), \mu_2(g_j))) | 1 \le i \le \mathrm{m}, 1 \le j \le \mathrm{n}\}.$$

- The coalescence-Max operation. Coalescence-Max on two equivalent objects o_1 and o_2, denoted by ©, is defined as $o = o_1$ © o_2, in which $o \approx o_1 \approx o_2$ and

$$o\ (fpM) = \{\max(\max_{i,j}(f_i, g_j), \min_{i,j}(\mu_1(f_i), \mu_2(g_j))) | 1 \leq i \leq m, 1 \leq j \leq n\}.$$

Then object o is a probabilistic object with a fuzzy measure o (*fpM*), which is created by coalescing two equivalent objects o_1 and o_2. In other words, o_1 and o_2 are two equivalent objects with degree of o (*fpM*).

Example: Suppose we two equivalent objects o_1 and o_2, where $o_1 \approx o_2$, o_1 (*fpM*) = {(0.2, 0.3), (0.3, 0.5)} and o_2 (*fpM*) = {(0.1, 0.4), (0.2, 0.5)}. Applying the operations defined above, we have

- for $o = o_1 \oplus o_2$, o (*fpM*) = {max ((min (1, 0.2 + 0.1), min (0.3, 0.4)), (min (1, 0.2 + 0.2), min (0.3, 0.5)), (min (1, 0.3 + 0.1), min (0.5, 0.4)), (min (1, 0.3 + 0.2), min (0.5, 0.5)))} = {max ((0.3, 0.3), (0.4, 0.3), (0.4, 0.4), (0.5, 0.5))} = {(0.3, 0.3), (0.4, 0.4), (0.5, 0.5)},
- for $o = o_1 \ominus o_2$, o (*fpM*) = {max ((max (0.2 – 0.1, 0), min (0.3, 0.4)), (max (0.2 – 0.2, 0), min (0.3, 0.5)), (max (0.3 – 0.1, 0), min (0.5, 0.4)), (max (0.3 – 0.2, 0), min (0.5, 0.5)))} = {max ((0.1, 0.3), (0.0, 0.3), (0.2, 0.4), (0.1, 0.5))} = {(0.1, 0.5), (0.2, 0.4)},
- for $o = o_1 \otimes o_2$, o (*fpM*) = {max (((0.2 × 0.1), min (0.3, 0.4)), ((0.2 × 0.2), min (0.3, 0.5)), ((0.3 × 0.1), min (0.5, 0.4)), ((0.3 × 0.2), min (0.5, 0.5)))} = {max ((0.02, 0.3), (0.04, 0.3), (0.03, 0.4), (0.06, 0.5))} = {(0.02, 0.3), (0.03, 0.4), (0.04, 0.3), (0.06, 0.5)}, and
- for $o = o_1 © o_2$, o (*fpM*) = {max ((max (0.2, 0.1), min (0.3, 0.4)), (max (0.2, 0.2), min (0.3, 0.5)), (max (0.3, 0.1), min (0.5, 0.4)), (max (0.3, 0.2), min (0.5, 0.5))} = {max ((0.2, 0.3), (0.2, 0.3), (0.3, 0.4), (0.3, 0.5))} = {(0.2, 0.3), (0.3, 0.5)}.

For two probabilistic objects with fuzzy measures, they may be semantically similar even if they are not equivalent. Here we introduce a notation of semantic similarity between two probabilistic objects with fuzzy measures. For two probabilistic classes with fuzzy measures, say c_1 and c_2, let c_1 contain attributes $\{a_1, a_2, \ldots, a_k, fpM_1\}$ and c_2 contain attributes $\{b_1, b_2, \ldots, b_l, fpM_2\}$, in which fpM_1 and fpM_2 are the fuzzy probabilistic attributes of c_1 and c_2, respectively. Let o_1 and o_2 be objects of class c_1 and c_2, respectively. Then for o_1 with $\{a_1, a_2, \ldots, a_k\}$ and o_2 with $\{b_1, b_2, \ldots, b_l\}$, the probability that o_1 and o_2 are semantically similar each other is first defined as follows.

$$SS'(o_1, o_2) = \sum_{i=1,2,\ldots,k}^{j=1,2,\ldots,l} |o_1(a_i) = o_2(b_j)| / (k + l - \sum_{i=1,2,\ldots,k}^{j=1,2,\ldots,l} |o_1(a_i) = o_2(b_j)|)$$

Here $\sum_{i=1,2,\ldots,k}^{j=1,2,\ldots,l} |o_1(a_i) = o_2(b_j)|$ means the numbers of attribute value pairs from o_1 and o_2 which are equivalent each other. Considering the values of the fuzzy probabilistic attributes of o_1 and o_2, the final probability that o_1 and o_2 are semantically similar each other is defined

$$\text{SS}(o_1, o_2) = \min(\text{SS}'(o_1, o_2), o_1(fpM_1), o_2(fpM_2))$$

The calculation of *min* (SS' (o_1, o_2), o_1 (fpM_1), o_2 (fpM_2)) follows the coalescence operation of the fuzzy probabilistic attribute values in o_1 © o_2 presented above.

4.2 Probabilistic Object-Class Relationship

In the probabilistic object-oriented databases with fuzzy measures, we may have crisp classes and crisp objects, and also we may have probabilistic classes and probabilistic objects with fuzzy measures. So we can identify four kinds of object-class relationships, which are

(a) the object-class relationship between a crisp object and a crisp class,
(b) the object-class relationship between a crisp object and a probabilistic class with fuzzy measure,
(c) the object-class relationship between a probabilistic object with fuzzy measure and a crisp class, and
(d) the object-class relationship between a probabilistic object with fuzzy measure and a probabilistic class with fuzzy measure.

The object-class relationships in (b), (c) and (d) above are called *fuzzy probabilistic object-class relationships*. Actually the case in (a) can be regarded as a special case of fuzzy probabilistic object-class relationships. Estimating the fuzzy probability degree of an object to a class is crucial for fuzzy probabilistic object-class relationship when classes are instantiated. Without loss of generality, in the following we investigate the object-class relationship between a probabilistic object with fuzzy measure and a probabilistic class with fuzzy measure.

In the classical object-oriented databases, an object belongs to a class if each attribute value of the object is included in the corresponding attribute domain of the class. Similarly, in order to determine if a probabilistic object with fuzzy measure belongs to a probabilistic class with fuzzy measure, it is necessary to determine if the attribute domain of the class includes the attribute value of the object with respect to each common attribute (i.e., non fuzzy probabilistic attribute).

Let *c* be a probabilistic class with fuzzy measure which contains attributes $\{a_1, a_2, \ldots, a_k, fpM\}$ and let *o* be an object on attribute set $\{a_1, a_2, \ldots, a_k, fpM\}$. In *c*, each attribute a_i ($1 \le i \le k$) is connected with a domain denoted *dom* (a_i). Then *o* belongs to *c*, that is, *o* is an object of *c*, if and only if $o(a_i) \in dom(a_i)$ ($1 \le i \le k$), where $a_i \ne fpM$. The probability that *o* belongs to *c* is related to *o* (*fpM*), the value of *o* on the fuzzy probabilistic attribute, and the degree that *o* (*fpM*) is included in *dom* (*fpM*), the domain of *fpM*. The inclusion degree of *o* (*fpM*) with respect to *dom* (*fpM*) is denoted μ (*dom* (*fpM*), *o* (*fpM*)). Here *o* (*fpM*) is a fuzzy value representing the probability of *o*, and *dom* (*fpM*) is the domain of attribute *fpM*. The calculation of μ (*dom* (*fpM*), *o* (*fpM*)) is investigated in [18] and then μ (*dom (fpM), o (fpM))* = ID *(dom (fpM)), o (fpM)))*. Finally the probability that *o* belongs to *c* is defined as follow.

$$\rho_c (o) = min (\mu (o (fpM), dom (fpM)), o (fpM))$$

The calculation of *min* (μ *(o (fpM), dom (fpM)), o (fpM))* follows the coalescence operation of the fuzzy probabilistic attribute values in $o_1 © o_2$ presented above.

Also it is possible that *c* is a probabilistic class with fuzzy measure which contains attributes $\{a_1, a_2, \ldots, a_k, fpM_1\}$ and *o* be an object on attribute set $\{b_1, b_2, \ldots, b_l, fpM_2\}$. Then without consideration of the fuzzy probabilistic attribute, the probability that *o* belongs to *c* is defined as follows.

$$\rho'_c (o) = \sum_{i=1,2,\ldots,k}^{j=1,2,\ldots,l} | o(b_j) \in dom(a_i) | / (k + l - \sum_{i=1,2,\ldots,k}^{j=1,2,\ldots,l} | o(b_j) \in dom(a_i) |)$$

Here $\sum_{i=1,2,\ldots,k}^{j=1,2,\ldots,l} | o(b_j) \in dom(a_i) |$ represents the number of attribute values of *o* which are included in the attribute domains of *c*. Considering the values of the fuzzy probabilistic attributes of *o*, the final degree that *o* belongs to *c* is defined as follows.

$$\rho_c (o) = min (\rho'_c (o), \mu (o (fpM_2), dom (fpM_1)), o (fpM_2))$$

The calculation of *min* $(\rho'_c (o), \mu (o (fpM_2), dom (fpM_1)), o (fpM_2))$ also follows the coalescence operation of the fuzzy probabilistic attribute values in $o_1 © o_2$ presented above.

4.3 *Probabilistic Subclass-Superclass Relationship*

In the object-oriented databases, a subclass is created from a superclass by means of inheriting some attributes and methods of the superclass, overriding some attributes and methods of the superclass, and defining some new attributes and methods. Then a subclass is the specialization of the superclass and any object which belongs to the subclass must belong to the superclass.

In the probabilistic object-oriented databases with fuzzy measures, classes may be probabilistic ones with fuzzy measures. A subclass created from a probabilistic class with fuzzy measure must be a probabilistic class with fuzzy measure. Then the subclass-superclass relationship is uncertain. In other words, a class is a subclass of another class with a probabilistic degree at this point, and the probabilistic degree may be imprecise and represented by a fuzzy set. In the following, we investigate how to determine if two probabilistic classes with fuzzy measures have the subclass-superclass relationship.

Viewed from the point of an extensional class, where a class is defined by the list of its object instances, a probabilistic class with fuzzy measure, say c_1 with fuzzy probabilistic attribute fpM_1, is a subclass of another probabilistic class with fuzzy measure, say c_2 with fuzzy probabilistic attribute fpM_2 if and only if

(a) for any object o, o must be an object of c_2 if o is an object of c_1, and
(b) for the probability that o is an object of c_1 (namely, o (fpM_1)) and the probability that o is an object of c_2 (namely, o (fpM_2)), the inclusion degree that o (fpM_1) is included by o (fpM_2) is greater than 0, i.e., μ (o (fpM_1), o (fpM_2)) > 0.

Then c_1 is the subclass of c_2 with a probability, which is the minimum in the probabilities to which these objects belong to c_1.

It can be seen that we evaluate probabilistic subclass-superclass relationship above by using the inclusion of objects to classes. It is clear that such an approach cannot be applied if no objects are available. It means that the approach discussed above cannot be used to assess if two probabilistic classes with fuzzy measures have a subclass-superclass relationship when the classes are intensional classes because they are defined by a set of attributes and their admissible values and no objects are available for the classes. At this point, it is needed for us to determine how the subclass is included by the superclass according to the inclusion of the attribute domains in the subclass with respect to the attribute domains in the superclass.

Formally let c_1 and c_2 be two probabilistic classes with attributes $\{a_1, a_2, \ldots, a_k, a_{k+1}, \ldots, a_m, fpM_1\}$ and $\{a_1, a_2, \ldots, a_k, a'_{k+1}, \ldots, a'_m, a_{m+1}, \ldots, a_n, fpM_2\}$, respectively. In c_2, attributes $a_1, a_2, \ldots$, and a_k are directly inherited from $a_1, a_2, \ldots$, and a_k in c_1, attributes $a'_{k+1}, \ldots$, and a'_m are overridden from $a_{k+1}, \ldots$, and a_m in c_1, and attributes $a_{m+1}, \ldots$, and a_n are unique to c_1.An attribute in c_1 or c_2, say a_i, has a domain, denoted dom (a_i). Let a_i be an attribute of c_1 and a_j be an attribute of c_2. Then c_1 is a subclass of c_2 if and only if

(a) $dom\ (a_i) = dom\ (a_j)$ when $1 \le i, j \le k$ and $i = j$, and
(b) $dom\ (a_i) \subseteq dom\ (a_j)$ when $k + 1 \le i, j \le m$ and $i = j$.

Considering that c_1 and c_2 are probabilistic classes with fuzzy measures, the subclass-superclass relationship between c_1 and c_2 is uncertain. The probability that c_1 is a subclass of c_2 should be the inclusion degree of the domain of fuzzy probabilistic attribute in c_1 (i.e., dom (fpM_1)) with respect to the domain of fuzzy probabilistic attribute in c_2 (i.e., dom (fpM_2)). The probability that c_1 is a subclass of c_2 is denoted μ (c_1, c_2) and is defined as follows.

$$\mu\ (c_1, c_2) = \mu\ (dom\ (fpM_1), dom\ (fpM_2))$$

In subclass-superclass hierarchies, it is possible that a class has multiple superclasses, called multiple inheritance of class. Suppose that more than one of the superclasses have common attributes. At this point, if it is not explicitly declared that from which superclass the subclass inherits the attribute, ambiguity may arise in multiple inheritance of class. Formally let class c be a subclass of class c_1 and class c_2. Assume that the attribute a_i in c_1, denoted by a_i (c_1), is common to the attribute a_i in c_2, denoted by a_i (c_2). If *dom* (a_i (c_1)) and *dom* (a_i (c_2)) are identical, there is not a conflict in the multiple inheritance hierarchy and c inherits attribute a_i directly. If *dom* (a_i (c_1)) and *dom* (a_i (c_2)) are different, however, a conflict in the multiple inheritance hierarchy occurs. At this moment, which one of a_i (c_1)) and a_i (c_2) is inherited by c is determined by the following: if μ (*dom* (a_i (c_1)), *dom* (a_i (c_2))) > μ (*dom* (a_i (c_2)), *dom* (a_i (c_1))), then a_i (c_1) is inherited by c, else a_i (c_2) is inherited by c.

Note that in the multiple inheritance hierarchy above, the subclass has different degrees with respect to different superclasses, not being the same as the situation in classical object-oriented database systems.

5 Formal Probabilistic Object-Oriented Database Model with Fuzzy Measures

Based on the discussion above, we have known that a class in the probabilistic object-oriented database model with fuzzy measures may be a probabilistic one with fuzzy measure. Accordingly, in the probabilistic object-oriented database model with fuzzy measures, an object belongs to a class with a probability represented by a fuzzy set, and a class is the subclass of another class with a probability represented by a fuzzy set also. In the object-oriented databases, the specification of a class includes the definition of *ISA* relationships, attributes and method implementations. In order to specify a probabilistic class with fuzzy measure, some additional definitions are needed. In addition to these common attributes, a new attribute named fuzzy probabilistic attribute and denoted *fpM* should be added into the class to indicate the probability degree to which an object belongs to the class. If the class is a probabilistic subclass, its superclasses and the probability degree that the class is the subclass of the superclass should be illustrated in the specification of the class.

Formally, the definition of a probabilistic class with fuzzy measure is shown as follows.

CLASS *class-name* WITH PROBABILITY *degree*
  INHERITS $superclass_1$ WITH PROBABILITY $degree_1$
  ...
  INHERITS $superclass_k$ WITH PROBABILITY $degree_k$
  ATTRIBUTES
    $Attribute_1$: DOMAIN dom_1: TYPE OF $type_1$
    ...
    $Attribute_m$: DOMAIN dom_m: TYPE OF $type_m$
    Fuzzy Probabilistic Attribute: FUZZY DOMAIN: TYPE OF *real*
  METHODS
    ...
END

For each common attribute in the class (i.e., non fuzzy probabilistic attribute), it corresponds to a data type, which may be a simple type such as integer, real, Boolean and string, or may be a complex type such as set type and object type. For fuzzy probabilistic attribute, its data type is a fuzzy type based real, which allows the representation of descriptive form of imprecise probabilistic information. Only fuzzy probabilistic attribute has fuzzy type and fuzzy probabilistic attribute is explicitly indicated in a class definition. In the class definition above, we declare only the base type (i.e., real) of fuzzy probabilistic attribute and the fuzzy domain. A fuzzy domain is a set of possibility distributions, fuzzy linguistic terms or membership functions (each fuzzy term is associated with a membership function).

6 Conclusion

It is required in modeling real-world problems and constructing intelligent systems to integrate different methodologies and techniques, and it has been the quest and focus of significant interdisciplinary research efforts. The advantages of such a hybrid system are that the strengths of its partners are combined and the weaknesses of its partners are complementary one another. The fuzzy probabilistic object-oriented database model provides a flexible database model to model hybrid imprecise and uncertain information as well as complex objects. In this chapter, a probabilistic object-oriented database model is introduced, in which possibility distributions arise at the level of objects as probability measures. Such an extended object-oriented database model can be applied for modeling stochastic events which probabilities are represented by possibility distributions. Based on the extended object-oriented database model and the fuzzy probabilistic constraint, we discuss several important semantic relationships between object and object, object and class, and subclass and superclass. The strategies and approaches for merging equivalent objects are discussed. A probabilistic object-oriented database model with fuzzy measures is proposed in this chapter.

Hybrid imprecise and uncertain information generally has multiple and complicated characteristics. In the fuzzy multidatabase systems, for example, the integrated databases are consisted of such tuples that are connected with probability measures and which attribute values are fuzzy [19, 31]. In future work, we will investigate the object-oriented database model which may contain other hybrid imprecise and uncertain information. In particular, we will focus on a fuzzy and probabilistic object-oriented database model, in which the attribute values may be fuzzy ones instead of crisp ones in this chapter.

Acknowledgment. This work was supported by the *National Natural Science Foundation of China* (60873010 & 61073139) and in part by the *Program for New Century Excellent Talents in University* (NCET- 05-0288).

References

1. Baldwin, J.M., Lawry, J., Martin, T.P.: A note on probability/possibility consistency for fuzzy events. In: Proceedings of the 6th International Conference on Information Processing and Management of Uncertainty in Knowledge-Based Systems, Granada, Spain, pp. 521–525 (July 1996)
2. Barbara, D., Garcia-Molina, H., Porter, D.: The management of probabilistic data. IEEE Transactions on Knowledge and Data Engineering 4(5), 487–502 (1992)
3. Buckles, B.P., Petry, F.E.: A fuzzy representation of data for relational databases. Fuzzy Sets and Systems 7(3), 213–226 (1982)
4. Cao, T.H., Nguyen, H.: Uncertain and fuzzy object bases: a data model and algebraic operations, International Journal of Uncertainty. Fuzziness and Knowledge-Based Systems 19(2), 275–305 (2011)
5. Cao, T.H., Rossiter, J.M.: A deductive probabilistic and fuzzy object-oriented database language. Fuzzy Sets and Systems 140(1), 129–150 (2003)
6. Cavallo, R., Pittarelli, M.: The theory of probabilistic databases. In: Proceedings of the 13th VLDB Conference, Brighton, England, pp. 71–81 (September 1987)
7. de Tré, G., de Caluwe, R., Prade, H.: Null values in fuzzy databases. Journal of Intelligent Information Systems 30(2), 93–114 (2008)
8. Dey, D., Sarkar, S.A.: Probabilistic relational model and algebra. ACM Transactions on Database Systems 21(3), 339–369 (1996)
9. Eiter, T., Lu, J.J., Lukasiewicz, T., Subrahmanian, V.S.: Probabilistic object bases. ACM Transactions on Database Systems 26(3), 264–312 (2001)
10. Haas, P.J., Suciu, D.: Special issue on uncertain and probabilistic databases. The VLDB Journal 18(5), 987–988 (2009)
11. Kornatzky, Y., Shimony, S.E.: A probabilistic object-oriented data model. Data & Knowledge Engineering 12(2), 143–166 (1994)
12. Lakshmanan, L.V.S., Leone, N., Ross, R., Subrahmanian, V.S.: ProbView: a flexible probabilistic database system. ACM Transactions on Database Systems 22(3), 419–469 (1997)
13. Ma, Z.M., Yan, L.: Updating extended possibility-based fuzzy relational databases. International Journal of Intelligent Systems 22(3), 237–258 (2007)
14. Ma, Z.M., Yan, L.: Fuzzy XML data modeling with the UML and relational data models. Data & Knowledge Engineering 63(3), 970–994 (2007)
15. Ma, Z.M., Yan, L.: A literature overview of fuzzy database models. Journal of Information Science and Engineering 24(1), 189–202 (2008)
16. Ma, Z.M., Yan, L.: A literature overview of fuzzy conceptual data modeling. Journal of Information Science and Engineering 26(2), 427–441 (2010)
17. Ma, Z.M., Zhang, F., Yan, L.: Fuzzy information modeling in UML class diagram and relational database models. Applied Soft Computing 11(6), 4236–4245 (2011)
18. Ma, Z.M., Zhang, W.J., Ma, W.Y.: Extending object-oriented databases for fuzzy information modeling. Information Systems 29(5), 421–435 (2004)
19. Ma, Z.M., Yan, L.: Conflicts and their resolutions in fuzzy relational multidatabases. International Journal of Uncertainty, Fuzziness and Knowledge-Based Systems 18(2), 169–195 (2010)
20. Nuray-Turan, R., Kalashnikov, D.V., Mehrotra, S., Yu, Y.M.: Attribute and object selection queries on objects with probabilistic attributes. ACM Transactions on Database Systems 37(1), 3 (2012)

21. Parsons, S.: Current approaches to handling imperfect information in data and knowledge bases. IEEE Transactions on Knowledge and Data Engineering 8(2), 353–372 (1996)
22. Pittarelli, M.: An algebra for probabilistic databases. IEEE Transactions on Knowledge and Data Engineering 6(2), 293–303 (1994)
23. Prade, H., Testemale, C.: Generalizing database relational algebra for the treatment of incomplete/uncertain information and vague queries. Information Sciences 34(2), 115–143 (1984)
24. Raju, K.V.S.V.N., Majumdar, A.K.: Fuzzy functional dependencies and lossless join decomposition of fuzzy relational database systems. ACM Transactions on Database Systems 13(2), 129–166 (1988)
25. Umano, M., Fukami, S.: Fuzzy relational algebra for possibility-distribution-fuzzy-relational model of fuzzy data. Journal of Intelligent Information Systems 3(1), 7–27 (1994)
26. Yan, L., Ma, Z.M.: Algebraic operations in fuzzy object-oriented databases. Information Systems Frontiers (2013), doi:10.1007/s10796-012-9359-8
27. Yang, Q., Zhang, W.N., Liu, C.W., Wu, J., Yu, C.T., Nakajima, H., Rishe, N.: Efficient processing of nested fuzzy SQL queries in a fuzzy database. IEEE Transactions on Knowledge and Data Engineering 13(6), 884–901 (2001)
28. Zadeh, L.A.: Fuzzy sets. Information and Control 8(3), 338–353 (1965)
29. Zadeh, L. A.: The concept of a linguistic variable and its application to approximate reasoning. Information Sciences 8 (3), 199–249, 8 (4), 301–357, 9(1): 43–80 (1975)
30. Zadeh, L.A.: Fuzzy sets as a basis for a theory of possibility. Fuzzy Sets and Systems 1(1), 3–28 (1978)
31. Zhang, W.N., Laun, E., Meng, W.Y.: A methodology of integrating fuzzy relational databases in a multidatabase system. In: Proceedings of the 5th International Conference on Database Systems for Advanced Applications, Melbourne, Australia, pp. 401–410 (April 1997)
32. Zimanyi, E.: Query evaluation in probabilistic relational databases. Theoretical Computer Science 171(1-2), 179–219 (1997)

Probabilistic XML: Models and Complexity

Benny Kimelfeld and Pierre Senellart

Abstract. Uncertainty in data naturally arises in various applications, such as data integration and Web information extraction. Probabilistic XML is one of the concepts that have been proposed to model and manage various kinds of uncertain data. In essence, a probabilistic XML document is a compact representation of a probability distribution over ordinary XML documents. Various models of probabilistic XML provide different languages, with various degrees of expressiveness, for such compact representations. Beyond representation, probabilistic XML systems are expected to support data management in a way that properly reflects the uncertainty. For instance, query evaluation entails probabilistic inference, and update operations need to properly change the entire probability space. Efficiently and effectively accomplishing data-management tasks in that manner is a major technical challenge. This chapter reviews the literature on probabilistic XML. Specifically, this chapter discusses the probabilistic XML models that have been proposed, and the complexity of query evaluation therein. Also discussed are other data-management tasks like updates and compression, as well as systemic and implementation aspects.

1 Introduction

Data managed by modern database applications are often *uncertain*. A few examples are the following. When information from different sources is conflicting,

---

Benny Kimelfeld
IBM Research–Almaden
San Jose, CA 95120, USA
e-mail: `kimelfeld@us.ibm.com`

Pierre Senellart
Institut Télécom; Télécom ParisTech; CNRS LTCI
Paris, France
e-mail: `pierre.senellart@telecom-paristech.fr`

Z. Ma & L. Yan (Eds.): *Advances in Probabilistic Databases*, STUDFUZZ 304, pp. 39–66.
DOI: 10.1007/978-3-642-37509-5_3

inconsistent, or simply presented in incompatible forms, the result of integrating these sources necessarily involves uncertainty as to which fact is correct or which is the best mapping to a global schema. When data result from automatic and imprecise tasks, such as information extraction, data mining, or computer vision, it is commonly annotated by a score representing the *confidence* of the system in the correctness of the data. When data are gathered from sensor networks, they come with the inherent imprecision in the measurement of sensors. Even when data is generated by humans, they are not necessarily certain: diagnostics of diseases stored in a hospital database are affected by the imprecision of human judgment (in addition to that of the diagnostics themselves). This ubiquity of uncertain data is all the truer when one deals with the World Wide Web, which is a heterogeneous collection of data that is constantly updated by individuals and automated processes.

Data uncertainty is often ignored, or modeled in a specific, per-application manner. This may be an unsatisfying solution in the long run, especially when the uncertainty needs to be retained throughout complex and potentially imprecise processing of the data. As an example, consider sensor data being gathered in a database, mined to extract interesting patterns, annotated by human experts, then integrated together with the result of other such analyses, independently made. Each of these steps, from the initial collection to the final integration, should be aware of the uncertain character of handled data; furthermore, each of these steps may even introduce further uncertainty. The goal of uncertain data management is to provide a unifying framework and a unifying system to handle the semantics of uncertainty, in the database itself. This goal is in line with the motivation behind DBMSs themselves, which were proposed in the 1970s as a uniform solution to the problem of managing data, while replacing previous systems that were tied to particular applications (e.g., accounting, cataloging, etc.).

Naturally, there are various ways to model uncertainty. Examples include representation of missing information (from SQL NULLs to more elaborate models of *incomplete data* [34]), fuzzy logic and fuzzy sets [29], and the Dempster-Shafer theory [70]. In this chapter, we consider *probabilistic* models that represent probability distributions over ordinary databases, and are based on the rich mathematical formalism of probability theory. Of course, quite a few real-life applications provide data that are probabilistic in nature. Examples include conditional random fields [46] (used in information extraction) and statistics-based tasks involved in natural-language processing [50]. But even when applications do not provide obvious probabilistic data, they often provide confidence scores that can be mapped to probability values.

Representing and Querying Probabilistic Databases

A *probabilistic database* is, conceptually, a probability space over ordinary databases, each of which is called a *possible world* [22]. In practice, such a probability space is obtained by introducing uncertainty about the value (or existence) of individual data items. If there are many such uncertain data items, then the number of possible worlds may be too large to manage or even to store. However, applications usually make assumptions on the correlation among the uncertain items (e.g., independence), and such assumptions typically allow for a substantially smaller (e.g., logarithmic-size) representation of the probability space. Hence, from a database point of view, the goal is to provide a proper language and underlying data model to specify and represent probabilistic databases in a *compact* manner (e.g., by building in the model assumptions of independence).

But a database is not just about storage. A central role of a database is to provide a clean, easy, and general way of accessing its data (while abstracting away from the actual implementation). In particular, a database supports a high-level language like SQL or XQuery. In the context of probabilistic databases, the correspondent of querying is that of finding *events* and *inferring* their probabilities. Hence, we would like the database not just to "store probabilities," but to actually understand their semantics and support inference tasks. The common realization of that [22, 31, 45, 53] is to allow the user to phrase ordinary queries (of the kind she would pose to an ordinary database), while the database associates each query answer with its computed probability (i.e., the probability that the answer holds true in a random possible world).

Finally, another central task of a database is to provide high performance for the operations it supports (e.g., query evaluation). This aspect is particularly challenging in the case of a probabilistic database, due to the magnitude of the actual probability space that such a database can (compactly) represent. As an example, for query evaluation (under the semantics mentioned in the previous paragraph), the baseline way of computing the probabilities is through the enumeration of all possible worlds, which is prohibitively intractable. Hence, we would like the operations to be performed on the compact representation itself rather than on the possible worlds. From the theoretical-complexity point of view, we require efficiency to be under the assumption that the input consists of the database in its compact form; in particular, "polynomial-time" is in the size of the compact representation, and not in that of the implied probability space. Not surprisingly, this requirement leads very quickly to computational hardness [22] (and sometimes even hardness of approximation [26, 45]). However, as we discuss throughout the chapter, in the case of probabilistic XML there are a few important settings where querying is tractable.

There is a rich literature on probabilistic relational databases [17, 21, 32, 35, 56, 61, 67]. In contrast, here we discuss *probabilistic XML* models, which represent probabilistic spaces over labeled trees. XML is naturally adapted to a number of applications where data is tree-like, including Web data or natural language parsing, and we give next specific examples of applications of probabilistic XML models. Later in this chapter (Section 6.3), we discuss the connection between probabilistic relational models and probabilistic XML models.

Probabilistic XML Applications

Following are concrete examples of applications or application areas where probabilistic XML is a natural data model and, moreover, the need to *query* probabilistic XML arises.

- **XML data integration.** Assume that a number of sources on the Web export XML information in potentially different schemas. To represent the result of the integration, we need a way to capture the uncertainty in the schema mappings, in deduplication, or in resolving conflicting information. This uncertainty can be characterized by probabilistic mappings [26] and probabilistic data integration rules [38, 39]. The outcome of the integration process can naturally be viewed as probabilistic XML (which is useful to query, update, and so on).
- **Web information extraction.** Extracting information from Web data means detecting, in a Web page, instances of concepts, or relations between these instances, based on the content or structure of these Web pages. A typical output is therefore a tree-like document, with local annotations about extracted information. Current extraction techniques, whether they are unsupervised or rely on training examples, are by nature imprecise, and several possible annotations might be produced for the same part of the Web page, with confidence scores. This is for instance the case with conditional random fields for XML [36] that produce probabilistic labels for part of the original HTML document; probabilistic XML is a natural way to represent that.
- **Natural language parsing.** Parsing natural language consists in building syntax trees out of sentences. This is an uncertain operation, because of the complexity of the natural language, and its inherent ambiguity. Indeed, some sentences like "I saw her duck" have several possible syntax trees. A parser will typically rely on statistics gathered from corpora to assign probabilities to the different possible parse trees of a sentence [50]. This probability space of parse trees can then be seen as probabilistic XML data [18].
- **Uncertainty in collaborative editing.** Consider users collaborating to edit documentation structured in sections, subsections, paragraphs and so on, as in the online encyclopedia Wikipedia. In an open environment, some of these contributions may be incorrect, or even spam and vandalism. If we have some way to estimate the trustworthiness of a contributor, we can represent each individual edit as an uncertain operation on a probabilistic XML document that represents the integration of all previous edits [1].
- **Probabilistic summaries of XML corpora.** Querying and mining a large corpus of XML documents (e.g., the content of the DBLP bibliography) can be time-consuming. If we are able to summarize this corpus as a compact probabilistic model [6], namely probabilistic XML, we can then use this model to get (approximations of) the result of querying or mining operations on the original corpus.

Organization

The remaining of this chapter is organized as follows. We first introduce the basic concepts, mainly XML, probabilistic XML, p-documents, and ProTDB as our main example of a concrete p-document model (Section 2). Next, we talk about querying probabilistic documents in general, and within ProTDB in particular (Section 3). We then review and discuss additional models (Section 4) and additional problems of interest (Section 5). Finally, we discuss practical aspects of probabilistic XML systems (Section 6) and conclude (Section 7).

As a complement to this chapter, we maintain an updated list of resources (especially, a hyperlinked bibliography) pertaining to probabilistic XML online at `http://www.probabilistic-xml.org/`.

2 Probabilistic XML

In this section, we describe the formal setting of this chapter, and in particular give the formal definitions of our basic concepts: an (ordinary) XML document, a probabilistic XML space, and the p-document representation of probabilistic XML.

2.1 XML Documents

We assume an infinite set Σ of *labels*, where a label in Σ can represent an XML tag, an XML attribute, a textual value embedded within an XML element, or the value of an attribute. The assumption that Σ is infinite is done for the sake of complexity analysis. An *XML document* (or just *document* for short) is a (finite) directed and ordered tree, where each node has a label from Σ. The label of a document node v is denoted by $\mathrm{label}(v)$. We denote by $\mathbf{D}_\Sigma$ the (infinite) set of all documents.

As an example, the bottom part of Figure 1 shows a document d. In this figure, as well as in other figures, labels that represent textual values (e.g., "*car financing*") are written in italic font, as opposed to labels that represent tags (e.g., "title"), which are written in normal font. Note that the direction of edges is not explicitly specified, and is assumed to be downward. Similarly, order among siblings is assumed to be left-to-right.

2.2 px-Spaces

A *probabilistic XML space*, abbreviated *px-space*, is a probability space over documents. Although we will briefly discuss *continuous* px-spaces (Section 4.5), our focus is mainly on *discrete* px-spaces. So, unless stated otherwise, we will implicitly assume that a px-space is discrete. Specifically, we view a px-space as a pair $\mathscr{X} = (D, p)$, where D is a finite or countably infinite set of documents, and $p : D \to [0,1]$ is a *probability function* satisfying $\sum_{d \in D} p(d) = 1$. The *support* of a px-space $\mathscr{X} = (D, p)$ is the set of documents $d \in D$, such that $p(d) > 0$. We say that the px-space $\mathscr{X}$ is *finite* if $\mathscr{X}$ has a finite support; otherwise, $\mathscr{X}$ is *infinite*.

When there is no risk of ambiguity, we may abuse our notation and identify a px-space $\mathscr{X}$ by the random variable that gets a document chosen according to the distribution of $\mathscr{X}$. So, for example, if $\mathscr{X} = (D, p)$ and d is a document, then $\Pr(\mathscr{X} = d)$ (in words, *the probability that* $\mathscr{X}$ *is equal to* d) is $p(d)$ if $d \in D$, and 0 otherwise.

2.3 p-Documents

A px-space is encoded by means of a *compact representation*. Later in this chapter, we will discuss the plethora of representation models proposed and studied in the literature. The basic notion underlying most of those models is that of a *p-document* [4, 43].

Formally, a p-document is a tree $\mathscr{P}$ that is similar to an XML document, except that $\mathscr{P}$ has a distinguished set of *distributional nodes* in addition to the *ordinary nodes* (that have labels from Σ). The ordinary nodes of $\mathscr{P}$ may belong to documents in the encoded px-space. Distributional nodes, on the other hand, are used only for defining the probabilistic process that generates random documents (but they do not actually occur in those documents). As an example, Figure 1 shows a p-document $\mathscr{P}$, where the distributional nodes are the ones represented by boxes with rounded corners (and denoted by v_1, v_2, and so on). The words ind and mux inside those boxes will be discussed later. Each distributional node specifies a probability distribution over subsets of its children; later on, we will define several *types* of distributional nodes (like ind and mux), where each type defines the way these distributions are encoded. In the probabilistic process that generates a random document, a distributional node randomly chooses a subset of its children according to the distribution specified for that node. The root and leaves of a p-document are required to be ordinary nodes.

Next, we describe the px-space (D, p) defined by a p-document $\mathscr{P}$ by specifying a sampling process that generates a random document. Note that such a process well defines the px-space (D, p) as follows: D consists of all the documents that can be produced in this process, and $p(d)$ (where $d \in D$) is the probability that d is obtained in this process.

The random document is generated by the p-document $\mathscr{P}$ in two steps. First, each distributional node of $\mathscr{P}$ randomly chooses a subset of its children. Note that the choices of different nodes are not necessarily probabilistically independent. All the unchosen children and their descendants (even descendants that have been chosen by their own parents) are deleted. The second step removes all the distributional nodes. If an ordinary node u remains, but its parent is removed, then the new parent of u is the lowest ordinary node v of $\mathscr{P}$, such that v is a proper ancestor of u. Note that two different applications of the first step may result in the same random document generated (and for further discussion on that, see [43]).

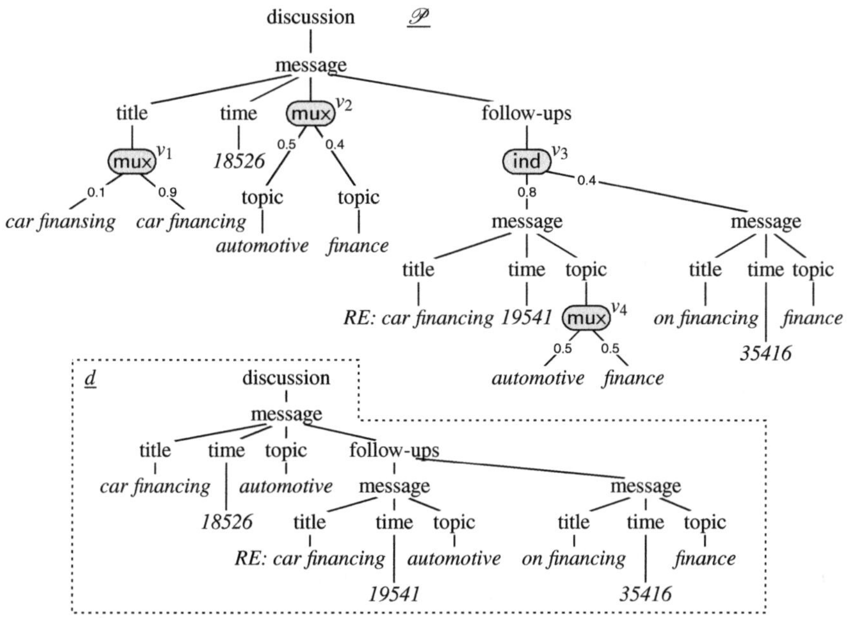

Fig. 1 A p-document $\mathscr{P}$ in $\mathsf{PrXML}^{\{\mathsf{ind},\mathsf{mux}\}}$ (top) and a sample document d of $\mathscr{P}$ (bottom)

A Concrete Model: ProTDB

We now construct a concrete model of p-documents, namely, the ProTDB model [53]. For that, we define two types of distributional nodes. Recall that when defining a type of distributional nodes, we need to specify the encoding and meaning of the random process in which a distributional node of that type selects children. In Section 4, we will define additional types of distributional nodes (hence, additional concrete models).

A ProTDB document has two types of distributional nodes.

- ind: A distributional node v of type ind specifies for each child w, the probability of choosing w. This choice is independent of the other choices of children, of either v or other distributional nodes in the p-document.
- mux: A distributional node v of type mux chooses at most one child w (that is, different children are mutually exclusive, hence the name mux) with a specified probability for w. We require the sum of probabilities along the children of v to be at most 1; the complement of this sum of probabilities is the probability that v chooses none of its children.

Example 1. The top part of Figure 1 shows a ProTDB p-document $\mathscr{P}$. The type of each distributional node is written in the corresponding box. For instance, node v_1 is a distributional node of type mux; as shown by the numbers on its outgoing edges, v_1 chooses its left child and right child with probability 0.1 and 0.9, respectively. Note that the mux node v_2 chooses none of its children with probability 0.1

($= 1 - 0.4 - 0.5$). Finally, observe that the ind node v_3 makes independent choices about its two children; for example, it chooses *just* the left child with probability $0.8 \times (1-0.4)$, both children with probability 0.8×0.4, and no children at all with probability $(1-0.8) \times (1-0.4)$.

In the bottom, Figure 1 shows a sample document d of $\mathscr{P}$. Let us now compute the probability of d. For d to be produced, the following independent events should take place:

- v_1 chooses its right child. This event occurs with probability 0.9.
- v_2 chooses its left child. This event occurs with probability 0.5.
- v_3 chooses both of its children. This event occurs with probability $0.8 \times 0.4 = 0.32$.
- v_4 chooses its right child. This event occurs with probability 0.5.

Hence, the probability of d is given by

$$\Pr(\mathscr{P} = d) = 0.9 \times 0.5 \times 0.32 \times 0.5 = 0.072\,. \qquad \square$$

We follow the conventional notation [4] that, given k types type_1, type_2, ..., type_k of distributional nodes (such as ind, mux, and the types that we define later), $\mathsf{PrXML}^{\{\mathrm{type}_1,\mathrm{type}_2,\dots,\mathrm{type}_k\}}$ denotes the model of p-documents that use distributional nodes only among type_1, type_2, ..., type_k. Hence, under this notation ProTDB is the model $\mathsf{PrXML}^{\{\mathsf{ind},\mathsf{mux}\}}$ (and for the p-document $\mathscr{P}$ of Figure 1 we have $\mathscr{P} \in \mathsf{PrXML}^{\{\mathsf{ind},\mathsf{mux}\}}$). Observe that $\mathsf{PrXML}^{\{\mathsf{ind},\mathsf{mux}\}}$ strictly contains $\mathsf{PrXML}^{\{\mathsf{ind}\}}$, $\mathsf{PrXML}^{\{\mathsf{mux}\}}$ and $\mathsf{PrXML}^{\{\}}$ (which is the set $\mathbf{D}_\Sigma$ of ordinary documents).

3 Query Evaluation

In this section, we discuss a central aspect in the management of probabilistic XML—query evaluation. In general, a query Q maps a document d to a value $Q(d)$ in some domain dom_Q; that is, a query is a function $Q : \mathbf{D}_\Sigma \to dom_Q$. As an example, in the case of a *Boolean* query, dom_Q is the set $\{\mathbf{true}, \mathbf{false}\}$; in that case we may write $d \models Q$ instead of $Q(d) = \mathbf{true}$ (and $d \not\models Q$ instead of $Q(d) = \mathbf{false}$). In the case of an *aggregate* query, dom_Q is usually the set $\mathbb{Q}$ of rational numbers. Later on, we discuss additional types of queries.

A px-space $\mathscr{X}$ and a query Q naturally define a probability distribution over dom_Q, where the probability of a value $a \in dom_Q$ is given by $\Pr(Q(\mathscr{X}) = a)$. We usually follow the conventional semantics [22] that, when evaluating Q over $\mathscr{X}$, the output represents that distribution. For example, if Q is a Boolean query, then the goal is to compute the number $\Pr(\mathscr{X} \models Q)$.

3.1 Query Languages

We now describe the languages of queries that capture the focus of this chapter: tree-pattern queries, monadic second-order queries, and aggregate queries.

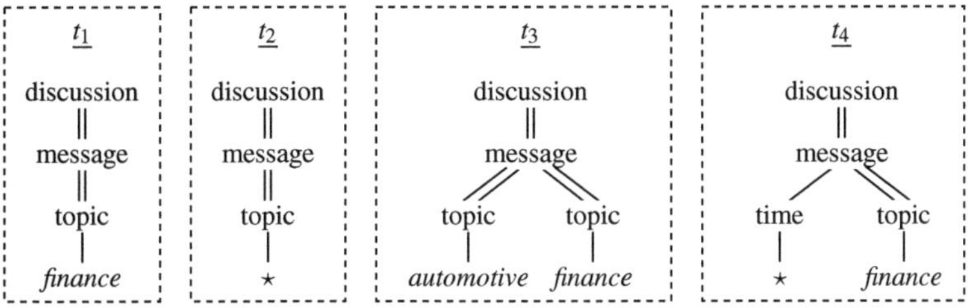

Fig. 2 Tree patterns

3.1.1 Tree-Pattern Queries

Tree-pattern queries (a.k.a. *twig queries* [9, 12]), or just *tree patterns* for short, correspond to the navigational fragment of XPath restricted to child and descendant edges. Specifically, a tree pattern is a Boolean query that is represented by a tree t with *child* and *descendant* edges. In our figures, child and descendant edges are depicted by single and double lines, respectively. Each node of the tree t is labeled with either a label of Σ or with the special *wildcard* symbol $\star$ (and we assume that $\star \notin \Sigma$). A *match* of a tree pattern t in a document d is a mapping μ from the nodes of t to those of d, such that μ maps root to root, child edges to edges, and descendant edges to paths (with at least one edge); furthermore, μ preserves labels of Σ, that is, for a node v of t, if $\text{label}(v) \neq \star$ then $\text{label}(v) = \text{label}(\mu(v))$. Note that a tree pattern ignores the order among siblings in a document. (Queries that take sibling order into account will be discussed in the next section.)

Example 2. Four tree patterns, $t_1, \ldots, t_4$, are shown in Figure 2. Child and descendant edges are represented by single and double lines, respectively. As in documents (and p-documents), edges are implicitly directed top down. As specific examples, let us consider the patterns t_1 and t_4. The pattern t_1 says that some message in the document has a topic descendant (where this topic can be that of the message or of a descendant message) with a child *finance*. The pattern t_4 is the same, except that it also requires the message to have a time child, and the time child to have a child (any child, as indicated by $\star$) of its own. □

Tree patterns are often used not just as Boolean queries, but also as queries that produce tuples of nodes (or tuples of labels). Informally speaking, these tuples are obtained by projecting the matches to a selected sequence of nodes of the tree pattern. For the sake of simplicity, here we restrict the discussion to the Boolean case. It is important to note, though, that under the standard notion of query evaluation for such queries[1] [22], evaluating a non-Boolean tree pattern reduces in polynomial time to evaluating a Boolean one [45].

---

[1] Under this notion, the output consists of every possible result tuple $\mathbf{a}$ and its marginal probability $\Pr(\mathbf{a} \in Q(\mathscr{X}))$.

3.1.2 Monadic Second-Order Tree Logic (MSO)

A language that is far more expressive than tree patterns is that of *Monadic Second-Order tree logic* (MSO). A query in MSO is a Boolean formula over the document nodes. The vocabulary includes two binary relations over nodes x and y: "x is the parent of y," denoted $E(x,y)$, and "x is a following sibling of y," denoted $y < x$. For each label $\lambda \in \Sigma$, the vocabulary includes also the unary relation "λ is the label of x," denoted $\lambda(x)$. The formula is in first-order logic, extended with quantification over set variables. This quantification allows, among other things, to express conditions on the set of all ancestors or descendants of a node, or on that of all nodes following a given node in document order. For a more formal definition the reader is referred to the vast literature on MSO for trees (e.g., Neven and Schwentick [52]).

Example 3. For illustration, the following MSO query says that there is a message with two descendants that are consecutive sibling messages on the topic *finance*.

$$\begin{aligned}\exists x, y_1, y_2 [&\text{message}(x) \wedge \text{message}(y_1) \wedge \text{message}(y_2) \\ &\wedge \underline{\textit{descendant}}(x,y_1) \wedge \underline{\textit{descendant}}(x,y_2) \wedge \underline{\textit{next-sibling}}(y_1,y_2) \\ &\wedge \underline{\textit{finance-topic}}(y_1) \wedge \underline{\textit{finance-topic}}(y_2)]\end{aligned}$$

In the formula above, $\underline{\textit{descendant}}(x,y)$ is phrased in MSO as follows.

$$\forall S[S(x) \wedge \forall z_1, z_2 (S(z_1) \wedge E(z_1,z_2) \rightarrow S(z_2)) \rightarrow S(y)]$$

Similarly, $\underline{\textit{next-sibling}}(y_1,y_2)$ is given by $y_1 < y_2 \wedge \neg\exists z[y_1 < z < y_2]$. Finally, $\underline{\textit{finance-topic}}(y)$ is phrased in MSO as follows.

$$\exists z, w[E(y,z) \wedge E(z,w) \wedge \text{topic}(z) \wedge \textit{finance}(w)] \qquad \square$$

MSO queries are closely related to the notion of a *(bottom-up) nondeterministic tree automaton* (NTA). Specifically, every MSO query can be translated into an NTA, such that the documents that satisfy the MSO query are precisely those that are accepted by the NTA; conversely, every NTA can be similarly translated into an MSO query [23, 52, 62].

3.1.3 Join Queries

Both tree patterns and MSO queries can be extended by adding *value joins* that test whether two nodes (for tree patterns), or two first-order variables (for MSO), have the same label. Value joins are fairly commonly used in XPath[2]; for instance, they allow us to dereference identifiers used as foreign keys.

Example 4. The following query in MSO extended with the *same-label* predicate tests whether two messages that are descendant of each other have the same topic:

[2] The first version of the XPath language only supports a limited form of value joins, but this restriction is lifted in the latest version.

$$\exists x_1,x_2,x_3,y_1,y_2,y_3[\text{message}(x_1)\wedge\text{message}(y_1)\wedge\underline{\mathit{descendant}}(x_1,y_1)$$
$$\wedge E(x_1,x_2)\wedge\text{topic}(x_2)\wedge E(x_2,x_3)$$
$$\wedge E(y_1,y_2)\wedge\text{topic}(y_2)\wedge E(y_2,y_3)$$
$$\wedge\,\mathit{same\text{-}label}(x_3,y_3)]$$ □

3.1.4 Aggregate Queries

In this chapter, an *aggregate function* is a function α that takes as input a set V of document nodes, and returns as output a numerical (rational) number $\alpha(V)\in\mathbb{Q}$. Some of the aggregate functions we consider, like sum, need to assume that the label of a node is a number; to accommodate that, we fix a function *num* over the document nodes, such that $\mathit{num}(v)=\text{label}(v)$ if label(v) is a number, and otherwise, we arbitrarily determine $\mathit{num}(v)=0$. Specifically, we will discuss the following aggregate functions.

- Count: $\mathsf{count}(V)\stackrel{\text{def}}{=}|V|$.
- Count distinct: $\mathsf{countd}(V)\stackrel{\text{def}}{=}|\{\text{label}(v)\mid v\in V\}|$; that is, $\mathsf{countd}(V)$ is the number of distinct labels that occur in V, regardless of the multiplicity of these labels.
- Sum: $\mathsf{sum}(V)\stackrel{\text{def}}{=}\sum_{v\in V}\mathit{num}(v)$.
- Average: $\mathsf{avg}(V)\stackrel{\text{def}}{=}\mathsf{sum}(V)/|V|$; if V is empty, then $\mathsf{avg}(V)$ is *undefined*.
- Min/max: $\mathsf{min}(V)\stackrel{\text{def}}{=}\min_{v\in V}\mathit{num}(v)$, $\mathsf{max}(V)\stackrel{\text{def}}{=}\max_{v\in V}\mathit{num}(v)$.

An *aggregate query* applies an aggregate function to the set of nodes that is selected by another query (of a different type). Specifically, here we consider aggregate queries that we write as $\alpha\circ t[w]$, where α is an aggregate function, t is a tree pattern, and w is a node of t. The evaluation of $\alpha\circ t[w]$ over a document d results in the number $\alpha(V)$, where V is the set of nodes v of d, such that there exists a match μ of t in d with $\mu(w)=v$; that is:

$$\alpha\circ t[w](d)\stackrel{\text{def}}{=}\alpha\left(\{v\mid\mu(w)=v\text{ for some match }\mu\text{ of }t\text{ in }d\}\right)$$

Example 5. Consider the tree pattern t_2 of Figure 2, and let w be the wildcard (denoted $\star$) node. When applied to the document d of Figure 1, the query $\mathsf{count}\circ t_2[w]$ returns 3, which is the number of nodes with a "topic" parent. In contrast, $\mathsf{countd}\circ t_2[w](d)$ is 2, which is the number of *distinct* topics (i.e., distinct labels of nodes with a "topic" parent) in d.

As another example, consider the tree pattern t_4 of Figure 2 and, again, let w be the wildcard node. The query $\mathsf{min}\circ t_4[w]$ returns the earliest time of a message that has a descendant message on the topic *finance*; hence, $\mathsf{min}\circ t_4[w](d)=\mathit{18526}$. □

3.2 *Complexity for ProTDB*

Nierman and Jagadish [53] studied the evaluation of (non-Boolean) tree patterns without projection, and showed computability in polynomial time. Although projection leads to hardness in the relational probabilistic model [22], Kimelfeld et al. [45] showed that tree patterns with projection, and in particular Boolean tree patterns, can be evaluated in polynomial time in ProTDB [45]. Cohen et al. [20] extended this result to MSO queries. The main reason behind this tractability is that it is possible to evaluate queries directly over a ProTDB tree in a bottom-up manner, making use of the locality of both the p-document and the query. This can be done using dynamic programming for tree patterns [43], and through the computation of a product automaton of the query and the p-document in the MSO case [10].

Theorem 1. [20] *Let Q be an MSO query (e.g., a tree pattern). The problem "compute* $\Pr(\mathscr{P} \models Q)$ *given* $\mathscr{P} \in \mathsf{PrXML}^{\{\mathsf{ind},\mathsf{mux}\}}$*" is in polynomial time.*

Observe that Theorem 1 is phrased in terms of *data complexity* [65], which means that the query is held fixed. As mentioned by Kimelfeld et al. [45], the evaluation of tree patterns becomes intractable if the query is given as part of the input. Actually, it was shown [20, 45] that over ProTDB the evaluation of tree patterns, and even MSO queries, is *fixed-parameter tractable* (abbr. FPT) [24], which means that only the coefficients (rather than the degree) of the polynomial depend on the query[3] (hence, FPT is stronger than "polynomial data complexity"). Nevertheless, while for tree patterns this dependence is "merely" exponential, for general MSO queries this dependence is not any elementary function (unless $P \neq NP$), since that is already the case when the p-document is ordinary (deterministic) [28, 51].

Tractability (in terms of data complexity) is lost when tree patterns are extended with (value) joins [5]. This is not surprising, for the following reason. Tree patterns with joins over trees can simulate Conjunctive Queries (CQs) over relations. Moreover, tree patterns with joins over $\mathsf{PrXML}^{\{\mathsf{ind}\}}$ can simulate CQs over "tuple-independent" probabilistic relations [22]. But the evaluation of CQs over tuple-independent probabilistic databases can be intractable even for very simple (and small) CQs [22]. Interestingly, it has been shown that adding *any* (single) join to *any* tree pattern results in a query that is intractable, unless that query is equivalent to a join-free pattern [42].

Theorem 2. [42] *If Q is a tree pattern with a single join predicate, then one of the following holds.*

1. *Q is equivalent to a tree pattern (hence, can be evaluated in polynomial time).*
2. *The problem "compute* $\Pr(\mathscr{P} \models Q)$ *given* $\mathscr{P} \in \mathsf{PrXML}^{\{\mathsf{ind},\mathsf{mux}\}}$*" is #P-hard.*

Recall that #P is the class of functions that count the number of accepting paths of the input of an NP machine [64]; this class is highly intractable, since using an oracle to a #P-hard function one can solve in polynomial time every problem in the polynomial hierarchy [63].

Next, we discuss aggregate queries. Cohen et al. [19] showed that for the aggregate functions count, min, and max, the evaluation of the corresponding aggregate queries is in polynomial time for $\mathsf{PrXML}^{\{\mathsf{ind},\mathsf{mux}\}}$. (That result of Cohen et al. [19] is actually for a significantly broader class of queries, which they refer to as "constraints.") Note that, for these specific functions, the number of possible results

[3] For a formal definition of FPT the reader is referred to Flum and Grohe's book [27].

(numbers) q is polynomial in the size of the input p-documents; hence the evaluation of an aggregate query Q reduces (in polynomial time) to the evaluation of $\Pr(Q(\mathscr{P}) = q)$.

Theorem 3. [19] *Let Q be the aggregate query $\alpha \circ t[w]$. If α is either* count, min *or* max, *then the problem "compute $\Pr(Q(\mathscr{P}) = q)$ given $\mathscr{P} \in \mathsf{PrXML}^{\{\mathsf{ind},\mathsf{mux}\}}$ and $q \in \mathbb{Q}$" is in polynomial time.*

Note that an immediate consequence of Theorem 3 is that we can evaluate, in polynomial time, Boolean queries like $\mathsf{count} \circ t[w] > q$ (i.e., where equality is replaced with a different comparison operator). Unfortunately, this result does not extend to the aggregate functions countd, sum and avg.

Theorem 4. [5, 19] *For each α among* countd, sum *and* avg *there is an aggregate query $Q = \alpha \circ t[w]$, such that the problem "determine whether $\Pr(Q(\mathscr{P}) = q) > 0$ given $\mathscr{P} \in \mathsf{PrXML}^{\{\mathsf{ind},\mathsf{mux}\}}$ and $q \in \mathbb{Q}$" is NP-complete.*

A particularly interesting fact that is shown by Theorems 3 and 4 is that there is an inherent difference between the complexity of count and countd when it comes to query evaluation over $\mathsf{PrXML}^{\{\mathsf{ind},\mathsf{mux}\}}$.

4 Additional p-Documents and Extensions

We now discuss additional representation systems for probabilistic XML. Some of these systems are p-document models with additional kinds of distributional nodes, and other systems are extensions of the p-document concept. We discuss the expressive power of these representation systems, and the complexity of query answering.

4.1 Long-Distance Dependencies

The mux and ind distributional nodes encode *local* dependencies between nodes, in the sense that the presence of a node in the document depends just on the presence of its parent and (in the case of mux) its siblings. However, it is often desired to represent *long-distance* dependencies to capture correlations among nodes of arbitrary locations in the document tree. Towards that, we introduce new kinds of distributional nodes. Assume a finite set $\{e_1 \dots e_n\}$ of independent Boolean random variables (called *Boolean events*), and a probability $\Pr(e_i)$ for each of these e_i. We define two new kinds of distributional nodes:

- cie [2, 4]: A distributional node v of type cie specifies for each child w of v a *conjunction of independent events* e_k or their negation $\neg e_k$ (e.g., $e_2 \wedge \neg e_5 \wedge e_6$).
- fie [41]: A distributional node v of type fie specifies for each child w of v an arbitrary propositional formula on the e_is (e.g., $e_2 \vee (e_3 \wedge \neg e_7)$).

Recall from Section 2 that, to define the semantics of a type of distributional node, we need to specify how a random subset of children is chosen by a node of that type.

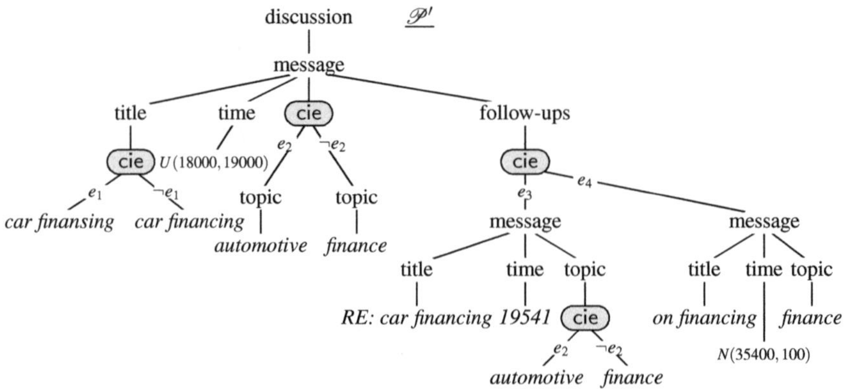

Fig. 3 A continuous p-document $\mathcal{P}'$ in $\text{PrXML}^{\{\text{cie}\}}$ with $\Pr(e_1) = 0.1$, $\Pr(e_2) = 0.5$, $\Pr(e_3) = 0.8$, $\Pr(e_4) = 0.4$

For cie and fie, the specification is as follows. At the beginning of the process, we draw a random truth assignment τ to the events $e_1, \dots, e_n$, independently of one another and according to the probabilities $\Pr(e_1), \dots, \Pr(e_n)$. Then, each distributional node selects the children that are annotated by a formula that evaluates to true under τ. (We then proceed to the second step, as described in Section 2.)

Example 6. An example p-document $\mathcal{P}'$ of $\text{PrXML}^{\{\text{cie}\}}$ is shown in Figure 3. Disregard for now the leaf nodes under "time" nodes (these nodes contain continuous distributions that will be discussed in Section 4.5). The p-document $\mathcal{P}'$ is somewhat similar to $\mathcal{P}$ of Figure 1: there is uncertainty in the title of the first message, in its topic, and in the existence of the two follow-up messages, which are independent of each other. However, there is also a fundamental difference. The topic of the first follow-up is correlated with that of the original message: either both are set to "*automotive*" or both are set to "*finance*." This reflects what a topic extraction system might do, if it has a global view of the whole discussion. □

We now look at the relative expressiveness and succinctness of p-documents defined with ind, mux, cie, and fie distributional nodes. In terms of expressiveness, $\text{PrXML}^{\{\text{ind,mux}\}}$, $\text{PrXML}^{\{\text{cie}\}}$, and $\text{PrXML}^{\{\text{fie}\}}$ are all able to represent all finite probability distributions over documents and are therefore equivalent [4] (as already noted, this is not the case for $\text{PrXML}^{\{\text{ind}\}}$, $\text{PrXML}^{\{\text{mux}\}}$ and, obviously, $\text{PrXML}^{\{\}}$). However, in terms of succinctness the picture is different: while there is a polynomial-time transformation of a $\text{PrXML}^{\{\text{ind,mux}\}}$ p-document into an equivalent $\text{PrXML}^{\{\text{cie}\}}$ p-document, the converse is not true [44]. Similarly, $\text{PrXML}^{\{\text{cie}\}}$ is a subset of $\text{PrXML}^{\{\text{fie}\}}$, but a transformation from $\text{PrXML}^{\{\text{fie}\}}$ into $\text{PrXML}^{\{\text{cie}\}}$ entails an inevitable exponential blowup [41].

The families $\text{PrXML}^{\{\text{cie}\}}$ and (a fortiori) $\text{PrXML}^{\{\text{fie}\}}$ are thus exponentially more succinct than ProTDB. However, this succinctness comes at a cost: query evaluation is now intractable. More precisely, *every* (Boolean) tree-pattern query is

#*P*-hard over $\mathsf{PrXML}^{\{\mathsf{cie}\}}$ (and $\mathsf{PrXML}^{\{\mathsf{fie}\}}$), except for some trivial cases [44, 45]. The situation in $\mathsf{PrXML}^{\{\mathsf{cie}\}}$ is essentially the same as that in $\mathsf{PrXML}^{\{\mathsf{fie}\}}$, although a few specific types of queries are tractable over $\mathsf{PrXML}^{\{\mathsf{cie}\}}$ and yet intractable over $\mathsf{PrXML}^{\{\mathsf{fie}\}}$: projection-free tree patterns with joins [58], and expected values for some types of aggregate queries [5, 7].

The intractability of querying p-documents with long-distance dependencies discussed above concerns the computation of the *exact* probability of a query. It makes sense to look also at approximation algorithms [45]. The simplest way to approximate query probability is by Monte-Carlo sampling: pick a random document, evaluate the query, and iterate. The approximated probability will then be the ratio of draws for which the probability evaluated to true. This approach yields a polynomial-time algorithm for obtaining an *additive* approximation of the query probability; that is, a number that is guaranteed, with high confidence, to be in the interval $[p-\varepsilon;p+\varepsilon]$ around the exact probability p. Using other means [37], in the case of tree patterns over $\mathsf{PrXML}^{\{\mathsf{cie}\}}$ it is also possible to obtain a (polynomial-time) *multiplicative* approximation (i.e., a number in the interval $[(1-\varepsilon)p,(1+\varepsilon)p]$) [45].

4.2 Conditional Models

As mentioned earlier, a central drawback in the ProTDB model (i.e., $\mathsf{PrXML}^{\{\mathsf{ind},\mathsf{mux}\}}$) and some other models proposed in the literature (e.g., [33]) is the assumption of probabilistic independence among probabilistic choices; in turn, this assumption is the key reason for the tractability of query evaluation [43]. However, even simple additional information about the database may give rise to intricate correlations. As a simple example, consider again the p-document in Figure 1. Even if we do not know the exact structure of the messages (hence, we use probabilistic rather than deterministic XML), it is likely that we know the total number of messages, and precisely (with no uncertainty involved). This new detail introduces dependency among the children of v_3, since now a random world cannot have too many (or too few) messages altogether. A more intricate statement can be the fact that at least 90% of the messages with the topic *automotive* have one or more *automotive* follow-ups; note that this statement implies correlation between the distributional nodes v_2 and v_4.

To incorporate such additional information, Cohen et al. [19] suggested to specify *constraints* in addition to the p-document. They presented a language for specifying constrains that may involve aggregate functions (e.g., "the total number of messages is 392," and "at least 80% of the messages have follow-ups").[4] Formally, a *Probabilistic XML Database* (PXDB) is a pair $(\mathscr{P},\mathscr{C})$, where $\mathscr{P}$ is a p-document and $\mathscr{C}$ is a set of *constraints*. The px-space that is defined by a PXDB $(\mathscr{P},\mathscr{C})$ is the sub-space of $\mathscr{P}$ conditioned on the satisfaction of each constraint of $\mathscr{C}$ (in other words, we restrict the px-space to the possible worlds that satisfy $\mathscr{C}$, and normalize the probabilities). Cohen et al. gave polynomial-time algorithms for various central tasks, such as *sampling* and *querying*, where their queries are tree patterns with

[4] For the precise specification of this language, see [19].

some aggregate functions (that include count, and min/max).[5] Similar tractability results have been shown for the case where both constraints and queries are phrased in MSO [20].

4.3 Recursive Markov Chains

In principle, p-documents provide means of representing arbitrary *finite* px-spaces. Some applications, however, require the ability to represent infinite sets of possible worlds. Consider again the example document of Figure 1; all such documents describing email discussions conform to the following schema, given as a DTD:

```
discussion: (message*)
   message: (title, time, topic?, follow-ups?)
follow-ups: (message*)
```

There are infinitely many documents conforming to this DTD, of arbitrarily large depth and width. In order to represent a discussion in which the number of messages and the structure of the discussion itself is fully uncertain, we need to be able to model, in a concise manner, infinite px-spaces.

The formalism of *recursive Markov chains* [25] is used for describing recursive probabilistic processes. Alternatives are described using a Markov chain, where each node in the chain can be a call to another (or the same) chain. This formalism naturally lends itself to the representation of potentially infinite px-spaces, as shown by Benedikt et al. [10]. In that work, Benedikt et al. study the tractability of MSO queries over px-spaces represented by recursive Markov chains (and restrictions thereof). In particular, recursive Markov chains that are *hierarchical* (i.e., when there are no cycles in the call graph) are tractable if we assume that all arithmetic operations have unit cost.[6] Hierarchical Markov chains can be seen as a generalization of p-documents defined with directed acyclic graphs instead of trees, a model introduced in [10, 20]. If we further restrict recursive Markov chains so that no Markov chain is called at two different positions (they are thus *tree-like*), we obtain a fully tractable model that generalizes $\mathsf{PrXML}^{\{\mathsf{mux},\mathsf{ind}\}}$ (and even more succinct models, e.g., $\mathsf{PrXML}^{\{\mathsf{exp}\}}$ [44]).

4.4 SCFGs

A *Context-Free Grammar* (CFG) specifies a process of producing parse trees for strings in a nondeterministic manner; indeed, a specific string may have multiple (even infinitely many) parse trees, since multiple production rules can be specified for a nonterminal. A *stochastic* (or *probabilistic*) *Context-Free Grammar* (SCFG) is similar to a CFG, except that the rules are augmented with probabilities; that is, the

[5] This language allows for nesting of queries, and the reader is referred to [19] for the exact details.

[6] Without this assumption, we lose tractability because the exact probability of a query may require exponentially many bits in the size of the representation.

production of a nonterminal becomes a probabilistic, rather than a nondeterministic, process.

When given a string, an SCFG implies a probability space over the possible parse trees of the string (where the probability of a parse tree corresponds to the confidence of the SCFG in that tree). Since this space comprises of labeled trees, we can view it as a (possibly infinite) px-space, on which we can evaluate XML queries (e.g., "find each noun phrase that forms an object for the verb *likes*"). Cohen and Kimelfeld [18] studied the problem of evaluating a tree-pattern query over the px-space that is represented by an SCFG and a string. In particular, they showed that this task is tractable for the class of *weakly linear SCFGs* (that generalizes popular normal forms like linear SCFGs, and Chomsky or Greibach normal forms). It follows from known results [25] that, in the general case, query probabilities do not have a polynomial-size bit representation, and can even be irrational.

4.5 Continuous Distributions

So far, all probabilistic XML models we have considered represent *discrete* probability distributions, where the uncertainty is either in the structure of the document or in the choice of a value from a finite collection of options. But some sources of uncertainty, such as the imprecision in sensor measurements, are essentially *continuous*. So, following [5, 7] we introduce the possibility of labeling leaves of p-documents with not only constant values, but *continuous probability distributions* of values (as usual, represented in some compact manner). For example, we might say that a given leaf represents a uniform distribution between two constants.

Example 7. Consider again the p-document $\mathscr{P}'$ of Figure 3. Two of the "time" nodes have for leaf a continuous distribution. The first one, $U(18000,19000)$ represents a uniform distribution in the interval $[18000;19000]$, which is adapted to the case when nothing else is known about the timestamp, perhaps because of a coarse granularity in the way the message metadata was displayed. The second distribution, $N(35400,100)$ is a Gaussian centered around 35400 and with a standard deviation of 100. Such a timestamp might arise from a known imprecision in the date of the computer system that produced the timestamp. One can check that the document d of Figure 1 is one of the possible worlds represented by $\mathscr{P}'$ (but of course, it has a zero probability due to the continuous distributions). □

Observe that we cannot use our current formalism of a px-space to define the semantics of a p-document with continuous values, since our px-space is discrete, and in particular, is defined by means of a probability of each possible world. Nevertheless, px-spaces can be properly extended to a continuous version by constructing a σ-algebra of sets of possible worlds, and define a probability measure over this σ-algebra, as done by Abiteboul et al. [7]. When this is done, we can investigate the complexity of query evaluation, as usual. Tree patterns are not of much interest in this case, because if a query node is matched against a node with continuous distribution, the probability of this match is usually zero. But of course, aggregate queries make sense. As shown by Abiteboul et al. [7], the tractability of aggregate queries with functions such as count, min, or max extends from (discrete) ProTDB to the continuous case, as long as the class of probability distributions present in the p-document can be efficiently convoluted, summed, integrated, and multiplied. This is for instance the case of distributions defined by piecewise polynomials, a generalization of uniform distributions.

5 Other Problems of Interest

In the previous sections, we discussed the task of query evaluation over different models of probabilistic XML. Here, we discuss additional tasks. Specifically, we address *updating* and *typing*, which are classical XML operations. We also discuss *compression*—the problem of finding a representation of a smaller size, and *top-k querying*—retrieving the most probable answers to a tree-pattern or a keyword-search query. Finally, we list additional tasks that are mostly left as open problems.

5.1 Updates

In update languages like *XUpdate* or the *XQuery Update Facility*, the specification of update operations entail *locator queries* that indicate, as XPath or XQuery expressions, the locations where data are to be inserted, modified, or deleted. An elementary probabilistic update operation can thus be defined as consisting of a locator query, a specification of the operation to be performed at matched locations (e.g., a tree to be inserted), and a probability that the update should be performed (provided that the locator query matches); such an operation has been studied by Abiteboul et al. [4]. The semantics of updates is defined as for queries: the result of an update on a probabilistic database should be a representation of a probabilistic space obtained from the original probabilistic space by applying the update on every possible world. Again, we want to avoid the exponential enumeration of possible worlds and perform the update directly on the original probabilistic document. Updates are of particular interest since they can be seen as a fundamental mechanism for constructing a probabilistic XML document: a sequence of uncertain update operations applied to a deterministic XML document [1].

Limiting our study to ProTDB and models with long-distance dependencies, we observe the following tradeoff on update tractability [41], in terms of data complexity:

- The result of an update operation is computable in polynomial time over ProTDB for a restricted set of non-branching tree pattern queries (specifically, those without descendant edges or those whose locator query returns the node at the bottom of the tree pattern).
- In general, computing the result of an update operation over ProTDB is intractable.
- The result of an update operation is computable in polynomial time over the family $\mathsf{PrXML}^{\{\mathsf{fie}\}}$, for updates defined by tree-pattern queries with joins.

The reason for the tractability of updates in $\mathsf{PrXML}^{\{\mathsf{fie}\}}$ (while querying operations are hard) is that updates do not entail computation of probabilities; we just manipulate event formulas without computing their probabilities.

Updating probabilistic XML documents highlights the following issue in models different from ProTDB and $\mathsf{PrXML}^{\{\mathsf{fie}\}}$: the model may lack the property of being a *strong representation system* [3] for the query language used in locator queries; this means that it is impossible to represent the output of a query (or the result of an update based on this query language) in the model. This is the case for ProTDB extended with continuous value distributions, and the language of aggregate tree-pattern queries (or even tree-pattern queries with inequalities). To be able to apply updates on such probabilistic models, the challenge is to define generalizations of these models (and of the corresponding querying techniques) that are strong representation systems.

5.2 Typing

Typing an XML document, that is, testing whether the document is valid against a schema defined in some schema language (e.g., DTD), is another fundamental data-management problem in XML. Similarly to Boolean querying, typing a probabilistic XML document should return a probability, namely, the probability that a random document is valid. As shown by Cohen et al. [20], when the schema can be defined by a deterministic bottom-up tree automaton (which is the case for DTDs, disregarding for now keys and foreign keys), computing the probability that a ProTDB p-document is valid is in polynomial time in the size of both the p-document and the schema. Essentially, this computation is done by running the automaton over the p-document, maintaining on the way some data structures that allow us to compute the probability that a node has type q given the corresponding probabilities of its children. This result can be generalized in a number of ways. First, tractability extends to computing the probability of a fixed query (say, a tree pattern) in the probabilistic space that is restricted to only those worlds that are valid against a schema [20]. Second, the data model can be generalized to recursive Markov chains, and we basically have tractability in the same classes of recursive Markov chains where MSO query answering is tractable [10]. Third, adding constraints (such as keys and foreign keys) renders typing intractable, though it is still tractable to test whether the probability of being valid against a schema with constraints is exactly one [20].

5.3 Compression

A fundamental advantage of using probabilistic XML models, such as ProTDB, is their potential compactness in representing probabilistic spaces. Depending on the application, obtaining such a compact model might not be straightforward. The direct translation of a set of possible worlds with probabilities into a $\mathsf{PrXML}^{\{\mathsf{mux},\mathsf{ind}\}}$ document, for instance, simply enumerates all possible worlds as children of a mux node and has the same size as the original space. The *compression* or *simplification* problem [39] is to obtain, given a probabilistic XML document, another more compact document that defines the same px-space.

In ProTDB, a basic operation that can be used to simplify a p-document is to push distributional nodes down the tree whenever possible, merging ordinary nodes in the process [66]. Another direction is to apply regular XML compression techniques [13] to compress the probabilistic tree into a probabilistic DAG while retaining querying tractability (assuming unit-cost arithmetics), as discussed in Section 4.3. Veldman et al. [66] explored the combination of probabilistic XML simplification techniques with ordinary XML compression, demonstrating gain in the size of the representation.

5.4 Top-k Queries

Chang et al. [16] studied the problem of finding, in a probabilistic XML document, the *top-k* query answers, that is, the *k* answers with the highest probabilities (where *k* is a specified natural number). Their model of probabilistic XML is ProTDB, and as queries they considered projection-free path patterns. Another type of a top-*k* query arises in *keyword search*. Information retrieval by keyword search on probabilistic XML has been studied by Li et al. [47]. Specifically, they perform keyword search in the ProTDB model by adopting the notion of *Smallest Lower Common Ancestor* (SLCA) [69], which defines when an XML node constitutes an answer for a keyword-search query. More particularly, the problem they explore is that of finding the *k* nodes with the highest probabilities of being SLCAs in a random world.

5.5 Open Problems

We now discuss important open problems around management operations on probabilistic XML. Despite the existence of techniques for compressing ProTDB documents [66], we lack a good understanding on when compression is possible and whether it is possible to obtain an *optimal* representation (with respect to compactness) of a px-space, in ProTDB and other models. A fundamental problem related to this one concerns *equivalence* of probabilistic XML documents: decide whether two representations define the same px-space [39]. As shown in [57], this problem admits a randomized polynomial-time decision procedure for $\mathsf{PrXML}^{\{\mathsf{cie}\}}$ when p-documents are shallow, giving some hope of obtaining a more systematic procedure for minimizing the size of a p-document. Nevertheless, the exact complexities of the equivalence problem, of testing optimality, and of minimization itself, remain open problems.

Compressing a discrete px-space into a compact p-document is somewhat akin to the problem of XML schema inference from XML data [11]: in both cases, the goal is to obtain a compact model of a set of documents. There are two differences, however. First, an XML schema represents a set of XML documents, while a p-document represents a probabilistic distribution thereof. Second, it is assumed that XML schema inference generalizes the observation of the example documents and that some documents valid against the schema are not present in the original collection, while compression preserves the px-space. Relaxing this last assumption leads to the problem of *probabilistic schema inference*, that is, learning a probabilistic model, with potential generalization, for a corpus of XML documents. A first work in this direction is by Abiteboul et al. [6], where the skeleton of the schema is given, and probabilities are learned to optimize the likelihood of the corpus. Adapting XML schema inference techniques to directly generate probabilistic models would allow us to generalize any collection of XML documents as a probabilistic XML document.

The focus of most of the literature on probabilistic XML is on modeling and querying, while only little exploration has been done on other aspects of probabilistic XML management. One of the important aspects that deserve further exploration is that of *mining*, namely, discovering important patterns and trends (e.g., frequent items, correlations, summaries of data values, etc.) in probabilistic XML documents. Kharlamov and Senellart [40] discuss how some mining tasks can be answered using techniques of probabilistic XML querying. Nevertheless, it is to be explored whether other techniques (e.g., based on ordinary frequent itemset discovery) can provide more effective mining.

6 Practical Aspects

In this section, we discuss some practical aspects of probabilistic XML management. We first consider system architecture and indexing, and then elaborate on the practical challenges that remain to be tackled towards a full-fledged database-management system for probabilistic XML. (To the best of our knowledge, up to now only prototypical systems have been developed.)

6.1 System Architecture

The first question is that of the general architecture of a probabilistic XML system: should it be *(a)* built on top of a probabilistic relational database system, *(b)* based on a query-evaluation engine for ordinary XML, *or (c)* engineered from scratch to easily accommodate the existing algorithmic approaches for probabilistic XML? We overview these three approaches, pointing to preliminary work, and noting advantages and shortcomings of each.

Over a Probabilistic Relational Engine

Much effort has been put on building efficient systems for managing probabilistic relational data. These systems include Trio [67], MayBMS [32] and its query evaluator SPROUT [54] (in turn, these systems are usually built on top of an ordinary relational database engine). Leveraging these efforts to the probabilistic XML case makes sense, and basically amounts to encoding probabilistic XML data into probabilistic tables, and tree-pattern queries into conjunctive queries. This direction is explored by Hollander and van Keulen [31] with Trio, where feasibility is demonstrated for different kinds of XML-to-relation encodings. However, the relational queries that result from those encodings are of a specific form (e.g., inequalities are used to encode descendant queries) for which optimizations are not always available to the probabilistic relational engine.

Over of an XML Query Engine

Alternatively, it is possible to rely on native XML database systems (such as eXist[7] or MonetDB/XQuery[8]) to evaluate queries over probabilistic XML documents, delegating components such as indexing of document structure and query optimization to the underlying XML database engine. It requires either to modify the internals of the XML query evaluation engine to deal with probabilities, or to be able to rewrite queries over probabilistic XML documents as queries over ordinary documents. The latter approach is demonstrated by Senellart and Souihli [59]; there, tree-pattern queries with joins over p-documents of $\mathsf{PrXML}^{\{cie\}}$ are rewritten into XQuery queries that retrieve each query match, along with a propositional formula that represents the probability of the match. All XML processing is therefore handed out to the XQuery query engine, and the problem is reduced to probability evaluation of propositional formulas.

Independent Implementation

The previous two architectures do not make use of the specificities of probabilistic XML, and in particular, of the techniques that have been developed for querying probabilistic XML. An alternative is thus to design a probabilistic XML system around one or more of these techniques (e.g., bottom-up dynamic programming [43]), and thereby utilize the known algorithms at query time [44]. The downside of this approach is that existing algorithms are main-memory intensive. Furthermore, the implemented system is typically applicable to only a limited probabilistic model (e.g., [44] supports just ProTDB documents, though it should be possible to use a similar bottom-up approach for hierarchical Markov chains [10] and to support continuous distributions [7]), and to a limited class of queries (e.g., [44] supports just tree patterns, but it should also be possible to extend it to MSO by combining the algorithm of [20] and a toolkit such as Mona [30] for converting queries into tree automata).

6.2 *Indexing*

We now consider *indexing* as a mean of enhancing the efficiency of query evaluation over probabilistic XML. When a probabilistic XML system is implemented on top of an XML database system, we can rely on this system to properly index the tree structure and content. Still remaining is the question of how to provide efficient access to the probabilistic annotations.

[7] `http://exist.sourceforge.net/`

[8] `http://monetdb.project.cwi.nl/monetdb/XQuery/`

The PEPX system [48] proposes to index ProTDB documents in the following manner: instead of storing with each child of a mux or ind node the probability of being selected by its parent, store the marginal probability that the child exists. Coupled with indexing of the tree structure, it allows much more efficient processing of simple queries, since a single access suffices to retrieve the probability of a node, and accessing all ancestors of this node is not required. This approach has also been taken by Li et al. [49] who adapted the TwigStack algorithm [12] to the evaluation of projection-free patterns in a ProTDB document.

This is certainly not the last word on probabilistic XML indexing, though. An interesting direction would be to combine structure-based indexing with probability-based indexing. Conceivably, such an approach has the potential of enhancing the efficiency of finding the most probable answers [16] or answers with a probability above a specified threshold [43].

6.3 Remaining Challenges

We now highlight some of the challenges that remain on the way to implementing a full-fledged system for managing probabilistic XML.

We first discuss the choice of method for query evaluation. Depending on the data model in use, and depending on the query language, we have a variety of techniques, exact or approximate: bottom-up algorithm in the absence of long-distance correlations [43], naïve enumeration of all possible worlds, Monte-Carlo sampling, relative approximation [44], and so on. Each of these has specific particularities in terms of the range of query and data it can be applied to, its evaluation cost, and its approximation guarantee. Hence, it is likely that some methods are suitable in some cases and other methods are suitable in other cases. A system should have a wealth of evaluation techniques and algorithms, and should be able to make proper decisions on which technique to use for providing a quick and accurate result. For example, the system may be given precision boundaries, and it should then select the most efficient approximation that guarantees these boundaries. Alternatively, given a time budget, a system should be able to select an exact or approximation technique (as precise as possible) for performing query evaluation within that budget. This process can be carried out at the level of the whole query, or at the level of each sub-query. For instance, in some cases it may be beneficial to combine probabilities that are computed (for different parts of the query and/or the document) by deploying different techniques. This suggests relying on cost-based, optimizer-like, query planning where each implementation of a (sub-)query evaluation is associated with an estimated cost, of both time and approximation. First steps are highlighted in [60].

There is also a need for a deeper understanding of the connection between probabilistic XML and probabilistic relational data. This is obviously critical if one is to implement a probabilistic XML system on top of a probabilistic relational database; but, it is important in other architectures as well, for identifying techniques in the relational setting that carry over to the XML setting. The connection is not so straightforward. It is of course easy to encode trees into relations or relations into trees, but in both cases the encoding has special shapes: relations encoding trees are tree-like (with treewidth [55] one) and relations encoded as trees are shallow and have repetitive structure. Typical query languages are different, too: tree-pattern queries or MSO on one side, conjunctive queries or the relational algebra on the other. When trees are encoded into relations, tree-pattern queries become a particular kind of conjunctive queries, involving hierarchically structured self joins, a class for which it is not always possible to obtain efficient query plans over arbitrary databases [61]. Some results from the probabilistic XML setting (such as the bottom-up evaluation algorithm for ProTDB) have no clear counterpart in the relational world, and vice versa. A unifying view of both models would help in building systems for managing both probabilistic relational and XML data.

The last challenge we highlight is that of optimizing query evaluation by reusing computed answers of previous queries. This can be seen as a case of *query answering using views*, a problem that has been extensively studied in the deterministic XML setting [8, 15, 68]. There is little known on whether and how (materialized) views can be used for query answering in the probabilistic XML setting, though Cautis and Kharlamov [14] have made a preliminary study of the problem in the setting of ProTDB, where they show that the major challenge is not retrieving query answers, but computing their probabilities.

7 Conclusions

We reviewed the literature on probabilistic XML models, which are essentially representation systems for compactly encoding probability distributions over labeled trees. A variety of such representation systems have been proposed, and each provides a different trade-off between expressiveness and compactness on the one hand, and management complexity on the other hand. Specifically, ProTDB [53] and some of its extensions (e.g., ProTDB augmented with constraints or continuous distributions, and tree-like Markov chains) feature polynomial-time querying for a rich query language (MSO, or aggregate queries defined by tree-patterns). In contrast, query evaluation is intractable in other models such as $\mathsf{PrXML}^{\{\mathsf{fie}\}}$ (that allows for correlation among arbitrary sets of nodes) or arbitrary recursive Markov chains (that can represent spaces of unbounded tree height or tree width).

We mentioned various open problems throughout this chapter. Two of these deserve particular emphasis. First, the connection to probabilistic relational models needs better understanding, from both the theoretical viewpoint (e.g., what makes tree-pattern queries over ProTDB tractable, when they are encoded into relations?) and the practical viewpoint (e.g., can we build on a system such as Trio [67] or MayBMS [32] to effectively manage probabilistic XML data?). Second, further effort should be made to realize and demonstrate the ideal of using probabilistic XML databases, or probabilistic databases in general, to answer data needs of applications (rather than devising per-application solutions); we discussed some of the wide range of candidate applications in the introduction. We believe that the research of recent years, which is highly driven by the notable popularity of probabilistic databases in the database-research community, constitutes a significant progress towards this ideal, by significantly improving our understanding of probabilistic (XML) databases, by developing a plethora of algorithmic techniques, and by building prototype implementations thereof.

Acknowledgements. We are grateful to Evgeny Kharlamov for his helpful comments. This work was partially supported by the European Research Council grant Webdam (under FP7), grant agreement 226513.

References

[1] Abdessalem, T., Ba, M.L., Senellart, P.: A probabilistic XML merging tool. In: EDBT (2011)
[2] Abiteboul, S., Senellart, P.: Querying and updating probabilistic information in XML. In: Ioannidis, Y., Scholl, M.H., Schmidt, J.W., Matthes, F., Hatzopoulos, M., Böhm, K., Kemper, A., Grust, T., Böhm, C. (eds.) EDBT 2006. LNCS, vol. 3896, pp. 1059–1068. Springer, Heidelberg (2006)
[3] Abiteboul, S., Hull, R., Vianu, V.: Foundations of Databases. Addison-Wesley (1995)
[4] Abiteboul, S., Kimelfeld, B., Sagiv, Y., Senellart, P.: On the expressiveness of probabilistic XML models. VLDB Journal 18(5) (2009)
[5] Abiteboul, S., Chan, T.H.H., Kharlamov, E., Nutt, W., Senellart, P.: Aggregate queries for discrete and continuous probabilistic XML. In: ICDT (2010)
[6] Abiteboul, S., Amsterdamer, Y., Deutch, D., Milo, T., Senellart, P.: Optimal probabilistic generators for XML corpora. In: BDA (2011a)
[7] Abiteboul, S., Chan, T.H.H., Kharlamov, E., Nutt, W., Senellart, P.: Capturing continuous data and answering aggregate queries in probabilistic XML. ACM Transactions on Database Systems (2011b)
[8] Afrati, F.N., Chirkova, R., Gergatsoulis, M., Kimelfeld, B., Pavlaki, V., Sagiv, Y.: On rewriting XPath queries using views. In: EDBT. ACM International Conference Proceeding Series, vol. 360, pp. 168–179. ACM (2009)
[9] Amer-Yahia, S., Cho, S., Lakshmanan, L.V.S., Srivastava, D.: Minimization of tree pattern queries. In: SIGMOD (2001)
[10] Benedikt, M., Kharlamov, E., Olteanu, D., Senellart, P.: Probabilistic XML via Markov chains. Proceedings of the VLDB Endowement 3(1) (2010)
[11] Bex, G.J., Neven, F., Vansummeren, S.: Inferring XML schema definitions from XML data. In: VLDB (2007)
[12] Bruno, N., Koudas, N., Srivastava, D.: Holistic twig joins: optimal XML pattern matching. In: SIGMOD (2002)
[13] Buneman, P., Grohe, M., Koch, C.: Path queries on compressed XML. In: VLDB (2003)
[14] Cautis, B., Kharlamov, E.: Challenges for view-based query answering over probabilistic XML. In: AMW (2011)
[15] Cautis, B., Deutsch, A., Onose, N.: XPath rewriting using multiple views: Achieving completeness and efficiency. In: WebDB (2008)
[16] Chang, L., Yu, J.X., Qin, L.: Query ranking in probabilistic XML data. In: EDBT (2009)
[17] Cheng, R., Singh, S., Prabhakar, S.: U-DBMS: A database system for managing constantly-evolving data. In: VLDB (2005)

[18] Cohen, S., Kimelfeld, B.: Querying parse trees of stochastic context-free grammars. In: ICDT (2010)
[19] Cohen, S., Kimelfeld, B., Sagiv, Y.: Incorporating constraints in probabilistic XML. ACM Transactions on Database Systems 34(3) (2009)
[20] Cohen, S., Kimelfeld, B., Sagiv, Y.: Running tree automata on probabilistic XML. In: PODS (2009)
[21] Dalvi, N., Ré, C., Suciu, D.: Probabilistic databases: Diamonds in the dirt. Communications of the ACM 52(7) (2009)
[22] Dalvi, N.N., Suciu, D.: Efficient query evaluation on probabilistic databases. VLDB Journal 16(4) (2007)
[23] Doner, J.: Tree acceptors and some of their applications. Journal of Computer Systems and Science 4(5) (1970)
[24] Downey, R.G., Fellows, M.R.: Parameterized Complexity. Monographs in Computer Science. Springer (1999)
[25] Etessami, K., Yannakakis, M.: Recursive Markov chains, stochastic grammars, and monotone systems of nonlinear equations. Journal of the ACM 56(1) (2009)
[26] Fagin, R., Kimelfeld, B., Kolaitis, P.: Probabilistic data exchange. In: ICDT (2010)
[27] Flum, J., Grohe, M.: Parameterized Complexity Theory. In: Texts in Theoretical Computer Science. An EATCS Series. Springer (2006)
[28] Frick, M., Grohe, M.: The complexity of first-order and monadic second-order logic revisited. Annals of Pure and Applied Logic 130(1-3) (2004)
[29] Galindo, J., Urrutia, A., Piattini, M.: Fuzzy Databases: Modeling, Design And Implementation. IGI Global (2005)
[30] Henriksen, J., Jensen, J., Jørgensen, M., Klarlund, N., Paige, B., Rauhe, T., Sandholm, A.: Mona: Monadic second-order logic in practice. In: Brinksma, E., Steffen, B., Cleaveland, W.R., Larsen, K.G., Margaria, T. (eds.) TACAS 1995. LNCS, vol. 1019, pp. 89–110. Springer, Heidelberg (1995)
[31] Hollander, E., van Keulen, M.: Storing and querying probabilistic XML using a probabilistic relational DBMS. In: MUD (2010)
[32] Huang, J., Antova, L., Koch, C., Olteanu, D.: MayBMS: a probabilistic database management system. In: SIGMOD (2009)
[33] Hung, E., Getoor, L., Subrahmanian, V.S.: PXML: A probabilistic semistructured data model and algebra. In: ICDE (2003)
[34] Imieliński, T., Lipski Jr., W.: Incomplete information in relational databases. Journal of the ACM 31(4) (1984)
[35] Jampani, R., Xu, F., Wu, M., Perez, L.L., Jermaine, C.M., Haas, P.J.: MCDB: a Monte Carlo approach to managing uncertain data. In: SIGMOD (2008)
[36] Jousse, F., Gilleron, R., Tellier, I., Tommasi, M.: Conditional random fields for XML trees. In: ECML Workshop on Mining and Learning in Graphs (2006)
[37] Karp, R.M., Luby, M., Madras, N.: Monte-Carlo approximation algorithms for enumeration problems. Journal of Algorithms 10(3) (1989)
[38] van Keulen, M., de Keijzer, A.: Qualitative effects of knowledge rules and user feedback in probabilistic data integration. VLDB Journal 18(5) (2009)
[39] van Keulen, M., de Keijzer, A., Alink, W.: A probabilistic XML approach to data integration. In: ICDE (2005)
[40] Kharlamov, E., Senellart, P.: Modeling, querying, and mining uncertain XML data. In: Tagarelli, A. (ed.) XML Data Mining: Models, Methods, and Applications. IGI Global (2011)

[41] Kharlamov, E., Nutt, W., Senellart, P.: Updating probabilistic XML. In: EDBT/ICDT Workshop on Updates in XML (2010)

[42] Kharlamov, E., Nutt, W., Senellart, P.: Value joins are expensive over (probabilistic) XML. In: LID (2011)

[43] Kimelfeld, B., Sagiv, Y.: Matching twigs in probabilistic XML. In: VLDB (2007)

[44] Kimelfeld, B., Kosharovski, Y., Sagiv, Y.: Query efficiency in probabilistic XML models. In: SIGMOD (2008)

[45] Kimelfeld, B., Kosharovsky, Y., Sagiv, Y.: Query evaluation over probabilistic XML. VLDB Journal 18(5) (2009)

[46] Lafferty, J., McCallum, A., Pereira, F.: Conditional Random Fields: Probabilistic models for segmenting and labeling sequence data. In: ICML (2001)

[47] Li, J., Liu, C., Zhou, R., Wang, W.: Top-k keyword search over probabilistic XML data. In: ICDE (2011)

[48] Li, T., Shao, Q., Chen, Y.: PEPX: a query-friendly probabilistic XML database. In: MUD (2006)

[49] Li, Y., Wang, G., Xin, J., Zhang, E., Qiu, Z.: Holistically twig matching in probabilistic XML. In: ICDE (2009)

[50] Manning, C.D., Schutze, H.: Foundations of Statistical NLP. MIT Press (1999)

[51] Meyer, A.R.: Weak monadic second-order theory of successor is not elementary recursive. Logic Colloquium 453 (1975)

[52] Neven, F., Schwentick, T.: Query automata over finite trees. Theoretical Computer Science 275(1-2) (2002)

[53] Nierman, A., Jagadish, H.V.: ProTDB: Probabilistic data in XML. In: VLDB (2002)

[54] Olteanu, D., Huang, J., Koch, C.: SPROUT: Lazy vs. eager query plans for tuple-independent probabilistic databases. In: ICDE (2009)

[55] Robertson, N., Seymour, P.D.: Graph minors. III. Planar tree-width. Journal of Combinatorial Theory, Series B 36(1) (1984)

[56] Sen, P., Deshpande, A., Getoor, L.: PrDB: managing and exploiting rich correlations in probabilistic databases. VLDB Journal 18(5) (2009)

[57] Senellart, P.: Comprendre le Web caché. Understanding the hidden Web. PhD thesis, Université Paris XI (2007)

[58] Senellart, P., Abiteboul, S.: On the complexity of managing probabilistic XML data. In: PODS (2007)

[59] Senellart, P., Souihli, A.: ProApproX: A lightweight approximation query processor over probabilistic trees. In: SIGMOD (2011)

[60] Souihli, A.: Efficient query evaluation over probabilistic XML with long-distance dependencies. In: EDBT/ICDT PhD Workshop (2011)

[61] Suciu, D., Olteanu, D., Ré, C., Koch, C.: Probabilistic Databases. Morgan & Claypool (2011)

[62] Thatcher, J.W., Wright, J.B.: Generalized finite automata theory with an application to a decision problem of second-order logic. Mathematical Systems Theory 2(1) (1968)

[63] Toda, S., Ogiwara, M.: Counting classes are at least as hard as the polynomial-time hierarchy. SIAM Journal on Computing 21(2), 316–328 (1992)

[64] Valiant, L.G.: The complexity of computing the permanent. Theoretical Computer Science 8, 189–201 (1979)

[65] Vardi, M.Y.: The complexity of relational query languages (extended abstract). In: STOC (1982)

[66] Veldman, I., de Keijzer, A., van Keulen, M.: Compression of probabilistic XML documents. In: Godo, L., Pugliese, A. (eds.) SUM 2009. LNCS, vol. 5785, pp. 255–267. Springer, Heidelberg (2009)
[67] Widom, J.: Trio: A system for integrated management of data, accuracy, and lineage. In: CIDR (2005)
[68] Xu, W., Özsoyoglu, Z.: Rewriting XPath queries using materialized views. In: VLDB (2005)
[69] Xu, Y., Papakonstantinou, Y.: Efficient keyword search for smallest LCAs in XML databases. In: SIGMOD (2005)
[70] Zadeh, L.A.: A simple view of the Dempster-Shafer theory of evidence and its implication for the rule of combination. AI Magazine 7(2) (1986)

Uncertain Data: Representations, Query Processing, and Applications

Tingjian Ge, Alex Dekhtyar, and Judy Goldsmith

Abstract. Uncertain data is common in many emerging applications. In this chapter, we start by surveying a few applications in sensor networks, ubiquitous computing, and scientific databases that require managing uncertain and probabilistic data. We then present two approaches to meeting this requirement. In the first approach, we propose a rich treatment of probability distributions in the system, in particular the SPO framework and the SP-algebra. In the second approach, we stay closer to a traditional DBMS, extended with tuple probabilities or attribute probability distributions, and study the semantics and efficient processing of queries.

1 Probabilistic Databases and Their Applications

There is a wide range of emerging applications that produce uncertain data and demand new techniques to manage such data. The mature, industry-standard relational database management systems have a history of about 40 years, but they do not have the capability of managing uncertain or probabilistic data. The applications that are discussed in this chapter are mainly in the areas of sensor networks, ubiquitous computing, bioinformatics and scientific databases. There are many applications (often related to the Internet) that also fall in this domain, such as information extraction and information integration.

In this section, we present several applications where large collections of probabilistic data are acquired, stored, and used. We divide those applications into two categories: sensor networks and ubiquitous computing, and scientific databases.

Tingjian Ge
University of Massachusetts, Lowell, MA, USA
e-mail: ge@cs.uml.edu

Alex Dekhtyar
California Polytechnic State University, San Luis Obispo, CA, USA
e-mail: dekhtyar@calpoly.edu

Judy Goldsmith
University of Kentucky, Lexington, KY, USA
e-mail: goldsmith@cs.uky.edu

Z. Ma & L. Yan (Eds.): *Advances in Probabilistic Databases*, STUDFUZZ 304, pp. 67–108.
DOI: 10.1007/978-3-642-37509-5_4

Within sensor networks, we consider a person-based application, namely monitoring of individual soldiers' physical status in the field, and the larger-grained examples of monitoring and controlling traffic in a large city, and monitoring and controlling power use in a house or community. Within scientific databases, we look at storing and managing astronomical data, microarrays, and proteomics. These are meant to be illustrative examples of a few hot research areas that depend on intelligent handling of probabilistic data, not a comprehensive catalogue of probabilistic database applications.

1.1 Sensor Networks and Ubiquitous Computing

Sensor networks and ubiquitous computing are major trends in modern computing. For example, many smartphones provide location estimates using a variety of sensors, such as GPS, WiFi, and/or cellular triangulation. However, the correctness of the triangulation depends on the proximity of cell towers, and on the local interference. It is thus important to handle any new data management issues that arise from uncertain data. Let us look at some concrete application scenarios.

Soldier Physiologic Status Monitoring

In the Soldier Physiologic Status Monitoring application (Tatbul et al. 2004), sensors are embedded in a "smart uniform" that monitors key biological parameters to determine the physiological status of a soldier. Under the harsh environment of the battlefield, it is crucial that sufficient medical resources reach wounded soldiers in a timely manner. Sensors in a smart uniform monitor thermal signals, hydration levels, cognitive and life signs, and wound levels.

There are a few ways the soldier's physiological states can be estimated with different sensors and with different confidence. An algorithm computes an overall score indicating how much medical attention the soldier needs and how urgent his or her condition is.

Tuple ID	Soldier ID	Time	Location	Score for Medical Needs	Conf.
T1	1	10:50	(10, 20)	49	0.4
T2	2	10:49	(10, 19)	60	0.4
T3	3	10:51	(9, 25)	110	0.4
T4	2	10:50	(10, 19)	80	0.3
T5	4	10:49	(12, 7)	56	1.0
T6	3	10:50	(9, 25)	58	0.5
T7	2	10:50	(11, 19)	125	0.3

Fig. 1 A table generated by sensors monitoring soldiers' needs for medical attention. The Conf. (confidence) attribute is the probability of existence of the tuple.

In a central database, as shown in Figure 1, a table records the information sent out by the sensors in the soldiers' uniforms. Each tuple in the table is one estimate with its related confidence. Sensors might be broken in harsh environments. For high availability, there can be two sets of sensors in a soldier's uniform in case one of them breaks down or loses precision. When each sends out an estimate at about the same time and they are inconsistent, at most one of them can be correct (together they form a discrete distribution with the confidence indicating the weight of each). These estimates may differ considerably due to variations in sensors, possible lost network messages, and different estimation algorithms.

Dynamic Traffic Routing

A few projects in both academia and industry (e.g., the CarTel project at MIT[1] and a product at INRIX[2]) provide traffic-aware routing and traffic mitigation. The idea is that various sensing devices are embedded in the cars that travel on roads and highways in urban areas. Some of the sensors measure the location of the cars (e.g., GPS or WiFi (Thiagarajan et al. 2009)), while others estimate travel delays. A large number of sensors from many cars continuously send data to a server.

The server uses this data to give real-time route planning decisions to drivers (e.g., *what is the quickest way to travel from A to B right now*?). Compared to alternatives such as using a standard online map, the dynamic routing also considers real-time factors such as road accidents and rush-hour traffic.

Such a system uses the travel delays reported in a recent time window to infer the probability distribution of current delay at a road. Due to random factors, the best we can get is a distribution. Similar to Figure 1, a central database contains a relational table with a number of attributes such as *road_ID*, *road_length*, *date*, *time*, *speed_limit*, and *current_delay*. Here, the *current_delay* attribute of a road can be modeled as a probability distribution, which is learned from a set of delay readings sent out from that road. To answer a routing query as given above, the system may need to run a shortest path algorithm over road delays that are probability distributions.

Smart Energy Grids

There is increased interest in monitoring, predicting, and even generating energy from multiple sources. Consider a system that integrates gas, coal, nuclear power, solar, hydro, and wind power, that has chips on all electric devices that communicate with central power company servers, local servers, and weather stations. Power-intensive tasks, from washing machines to automated factories, could be set to run when the solar cells are likely to be charged, the windmills are likely to be active, or the demand for heating or air conditioning is expected to be low. South Korea is testing such a system (McDonald, 2011), as are other countries. Power agents are being developed in situ, and in the context of a Trading Agents Competition (Block et al. 2010).

A power agent needs to be able to reason about likely weather conditions and power demands over the immediate and near future. It needs to condition such reasoning on location, time of year, and recent power demands, and to know about the tasks it is assigned to schedule.

---

[1] `http://cartel.csail.mit.edu/doku.php`

[2] `http://www.inrix.com/`

1.2 Scientific Databases

Scientific observations are fundamentally uncertain. No measurement is exact. When a quantity is observed and measured, the outcome depends on the measuring system, the experimental procedure, the skill of the person conducting the experiment, the environment, and other effects. Even if the quantity were to be measured several times, in the same way and in the same circumstances, a different measured value would in general be obtained each time, assuming that the measuring system has sufficient resolution to distinguish between the values.

Furthermore, as observed by domain scientists (e.g., (Burton et al. 2009)), due to unknown complex factors, contemporary scientific problems (e.g., associations of genetic variants and chronic diseases) often demand vast sample sizes and it is much needed to synthesize data across many studies and to undertake a pooled analysis. Below, we will look at a few concrete examples.

Astronomy

In astronomy, observations of the objects and phenomena in the sky are typically associated with "error bars" that indicate the estimated Gaussian distributions for the values being observed. Let us look at one of the most popular astronomical dataset, the Sloan Digital Sky Survey (SDSS). SDSS is one of the most ambitious and influential surveys in the history of astronomy[3]. It covers more than a quarter of the sky and contains more than 930,000 galaxies and more than 120,000 quasars.

Function/Gene name	Fold difference
Cell migration/invasion	
matrix metaloprotease 10 (MMP10)	36.3± 3.73
matrix metaloprotease 13 (MMP13)	21.4±16.38
plasminogen activator inhibitor 2 type A (PAI2A)	118.2±30.21
urokinase plasminogen activator receptor 1 (uPAR-1/PLAUR)	8.7±1.38
carbonic anhydrase 2 (CAII)	23.0±18.90
Cell adhesion/cell-cell interaction	
CD44	25.2±9.31
parathyroid hormone-like peptide (PTHLH)	76.3±6.83
osteopontin (OPN/SPP1)	87.0±15.83
fibromodulin (FMOD)	− 17.2±0.00
cartilage link protein 1 (CRTL1)	− 12.5±2.33

Fig. 2 Fold differences of two function groups of genes (among many) as measured by a microarray experiment (Komatsu, et al. 2006)

In the SDSS dataset, objects can have positional attributes: *right ascension* (ra) and *declination* (dec) in the J2000 coordinate system. Besides these two attributes, there are another two attributes, *ra_error* and *dec_error*, which are error bars. They indicate that the right ascension (declination, respectively) attribute is a random variable that has a Gaussian distribution with a standard deviation *ra_error* (*dec_error*, respectively) and a mean *ra* (*dec*, respectively).

[3] http://www.sdss.org/

Microarrays

DNA microarray analysis has been one of the most widely used sources of genome-scale data in the life sciences. Microarray expression studies are producing massive quantities of gene expression and other functional genomics data, which promise to provide key insights into gene function and interactions within and across metabolic pathways.

Figure 2 shows a snippet of the result from a microarray experiment performed by a research group (Komatsu, et al. 2006). It shows the fold differences of genes under two function groups. Here, a fold difference value indicates the difference between the gene's expressed strength in a tissue sample (e.g., cancer cells) and that in a normal tissue being compared with. A positive (negative, resp.) value indicates that the gene is expressed more strongly (more weakly, resp.) in the tissue sample. Thus, scientists are interested in finding genes with large absolute fold differences, which are characteristic of the disease/tissue being studied. The ± range value (e.g., 3.73 in the first gene) is the standard deviation over a few repeated runs, each of which is called a replicate. We can see that the variance can be quite significant. Figure 2 only shows selected genes from two function groups among many.

PML_HUMAN Mass: 97455 **Score: 194** Expect: 1e-14 Matches: 15
Probable transcription factor PML (Tripartite motif-containing protein 19) (RING finger protein 71)

MURC_IDILO Mass: 52994 **Score: 51** Expect: 2 Matches: 5
UDP-N-acetylmuramate--L-alanine ligase (EC 6.3.2.8) (UDP-N-acetylmuramoyl-L-alanine synthetase) – I

DPO1_RICHE Mass: 104386 **Score: 50** Expect: 2.8 Matches: 6
DNA polymerase I (EC 2.7.7.7) (POL I) - Rickettsia helvetica

:

:

Fig. 3 Sample output from the Mascot software that displays the proteins found in a sample. The scores indicate the confidence of the detections.

Proteomics

Proteomics is the large-scale study of proteins, particularly their structures and functions. Mass spectrometry has become a powerful tool in protein analysis and the key technology in proteomics (Mann et al. 2001). Proteomics experimental results contain information such as what proteins are in a tissue (either with a certain disease or normal), and their abundance, etc. Due to the many technical constraints in mass spectrometry (Mann et al. 2001), the experimental results have significant uncertainty.

Figure 3 shows a piece of the sample output from the widely used Mascot software[4] using Peptide Mass Fingerprint. Each possible protein is associated with a score, indicating the confidence of the detection. This can become more complicated when a tissue sample contains multiple proteins. A scientist would be

[4] `http://www.matrixscience.com/`

interested in knowing the abundance of a protein in a tissue, etc. Such information is often compared between a tissue sample (e.g., cancer cells) and a control (i.e., normal cells).

We have seen motivating applications in domains that require data management systems to handle uncertain and probabilistic data. In the rest of this chapter, we focus on two approaches of probabilistic databases. In the first approach (Section 2), we propose a rich treatment of probability distributions as data, in particular the SPO framework and the SP-algebra. In the second approach (Sections 3 and 4), we stay closer to a traditional DBMS, extend it with tuple probabilities or attribute probability distributions, and then study the semantics and efficient processing of queries in this model.

2 Semistructured Probabilistic Database Management Systems

One of the most common data structures for probabilistic reasoning is the Bayesian network, or Bayes net (Pearl 1988). A Bayes net is a directed acyclic graph, where nodes represent random variables and edges represent dependencies; each node has a probability table for the associated variable, conditioned on the values of its parents in the graph.

Example. Consider the Soldier Physiologic Status Monitoring application discussed in Section 1.1. Let us assume that a soldier has three sensors that, at set time intervals, send information about his/her body temperature, oxygen levels and pulse back to the home base server. Based on information supplied by the three sensors, which we refer to as T (temperature), O (oxygen) and P (pulse), the monitors (human or automatic) at the home base make decisions concerning the current state of the soldier. In this simplified scenario, suppose there are four states that a soldier can be in: N(ormal), W(eak), A(gitated) or S(ick). The reports of each of the three sensors are discretized into two values: H(igh) and N(ormal), with specific values of body temperature (e.g., 101F), blood oxygen levels (e.g., 90%) and pulse (e.g, 84) serving as the boundary values between them. The current state of the soldier is determined based on the information obtained from these three sensors. This can be represented graphically in the form of a Bayes net shown in Figure 4.

To complete the Bayes net, we supply the conditional probability distribution for the random variable C(ondition) based on the value of random variables T(emperature), O(xygen) and P(ulse), and provide marginal probability distributions for the other three random variables. Table **1** shows the former, while Table 2 shows a joint probability distribution of T, O and P.

A software application for tracking the medical conditions of military personnel might have to operate with different conditional probability tables and different marginal probability distributions based on a variety of factors. For example, the distributions shown in Tables 1 and 2 may be based on performance of Army personnel in temperate, forest-covered hilly environment. A different set of probability distributions might cover mountainous or desert terrain and be constructed for

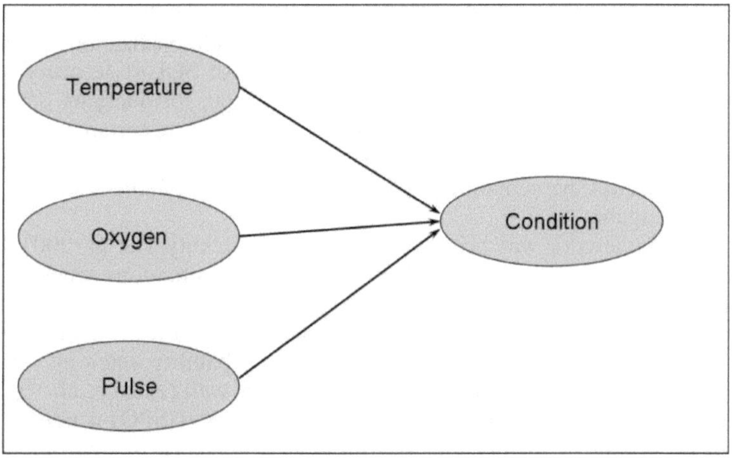

Fig. 4 A Bayes net for determining the physiological condition of a soldier in the field

other branches of the military (e.g., the Marine Corps) or specific military units. From joint probability distributions such as the one in Table 2, one can derive marginal probability distributions for subsets of parameters (e.g., Table 3 shows marginal probability distributions for (a) a pair of parameters T and P and (b) single parameter O) and can obtain conditional probability distributions (Table 3 (c) shows the distribution of T and O for personnel with high pulse rates).

Table 1 Conditional probability distribution for the Bayes net for determining the physiological condition of a soldier

T	O	P	Condition			
			Normal	Weak	Agitated	Sick
High	Normal	High	0.05	0.1	0.3	0.55
High	Normal	Normal	0.2	0.1	0.15	0.65
High	Low	High	0.05	0.05	0.1	0.8
High	Low	Normal	0.2	0.1	0.05	0.8
Normal	Normal	High	0.3	0.05	0.55	0.1
Normal	Normal	Normal	0.8	0.05	0.1	0.05
Normal	Low	High	0.2	0.2	0.45	0.15
Normal	Low	Normal	0.3	0.4	0.05	0.25

For decision-support software designed to work with this data, a data management mechanism is needed to deal with all such probability distributions. This is the underlying idea behind the Semistructured Probabilistic Objects (SPO) framework (Zhao et al. 2005). In this framework, diverse probability distributions, such as the ones depicted in Tables 1, 2, and 3 are stored as first-class database objects. A rich query algebra (the SP-algebra) is able to manipulate and retrieve the objects. The algebra incorporates traditional relational algebra operations of selection, projection, and

Cartesian product and join. It modifies their semantics to perform appropriate computations on the probability distributions, and adds a conditionalization operation that is unique to working with probability distributions. An SQL-style query language (SPOQL) has been implemented as a convenient syntax for querying databases of SPOs (Dekhtyar et al. 2006), though other implementations are certainly possible.

In what follows, we define the SPO framework formally, introduce the SP-algebra and discuss the semantics of its operations, and establish some key facts about the SP-algebra.

The SPO framework was originally introduced (Dekhtyar et al. 2001, Zhao et al. 2005) for exact (or "point") probabilities, i.e., for situations where the exact probabilities are known. However, it was observed that in many decision support applications, exact probabilities were not known. Rather, the probabilities of various situations/events were known to fall into probability intervals. The SPO framework was adapted to address such situations as well (Zhao et al. 2004). The notion of an Interval Semistructured Probabilistic Object (ISPO) is not too different than the notion of a SPO, but the Interval SP-algebra is significantly more complex (Zhao et. al 2003). We discuss this notion briefly at the end of Section 2.

Table 2 Joint marginal probability distribution of temperature, blood oxygen levels and pulse rate for the Bayes net for determining the physiological condition of a soldier.

T	O	P	*Prob*
High	Normal	High	0.02
High	Normal	Normal	0.01
High	Low	High	0.03
High	Low	Normal	0.04
Normal	Normal	High	0.25
Normal	Normal	Normal	0.4
Normal	Low	High	0.2
Normal	Low	Normal	0.05

Table 3 Probability distributions computable in the physiological condition monitoring scenario

(a)

T	P	*Prob*
High	High	0.05
High	Normal	0.05
Normal	High	0.45
Normal	Normal	0.45

(b)

O	*Prob*
Normal	0.68
Low	0.32

(c)

T	O	*Prob*
High	Normal	0.44
High	Low	0.06
Normal	Normal	0.5
Normal	Low	0.4
Prob(T,O|P=high)		

The term "semistructured" in the name of the framework was chosen for two reasons. The probability distributions stored in a single "relation" inside a Semistructured Probabilistic Database can have diverse structures and contain different "attributes". In addition to that, originally, XML was chosen as the representation syntax for SP objects (Dekhtyar et al. 2001). As XML representation is essentially syntactical in nature, we omit it from this narrative, and instead, concentrate on the semantics of the proposed frameworks.

2.1 Semistructured Probabilistic Objects

Consider a universe $\mathbf{V}$ of discrete random variables $\{v'_1,\ldots,v'_q\}$. With each random variable $v \in V$ we associate $dom(v)$, a finite set of its possible values. Given a set $V=\{v_1,\ldots,v_q\} \subseteq \mathbf{V}$, $dom(V)$ denotes $dom(v_1)x \ldots x\ dom(v_q)$.

Let $R=(A_1,\ldots,A_n)$ be a collection of regular relational attributes. For $A \in R$, let $dom(A)$ denote the domain of A. We define a semistructured schema $\underline{R}^*$ over R as a multiset of attributes from R. For example, if R = *{Terrain, MilitaryBranch, Conditions}* the following are valid semistructured schemas over R: R^*_1 = *{Terrain, MilitaryBranch}*; R^*_2 = *{Terrain, Conditions, Conditions}*; R^*_3 = *{Terrain, Terrain, Terrain}*.

Let $\mathbf{P}$ denote a probability space used in the framework to represent probabilities of different events. We present the framework over two different probability spaces. The first probability space $\mathbf{P}_{point}=[0,1]$, is the unit interval. Values from this interval are called *exact* or *point probabilities.* The Semistructured Probabilistic Object (SPO) framework introduced below uses this probability space. Another possibility, leading to an extended SPO framework (Zhao et al. 2004), is based on the probability space $\mathbf{P}_{int}=C[0,1]$: the set of all subintervals of the unit interval. A probability value from this space is called an *interval probability.* The general definition of a Semistructured Probabilistic Object given below applies for any probability space, however, the query algebra for each of the two frameworks is substantially different. We describe the query algebra, the SP-algebra, over the point probability space $\mathbf{P}_{poin}$ in Section 2.2, and briefly discuss the query algebra (Extended SP-algebra) for the interval probability space $\mathbf{P}_{int}$ at the end of Section 2.

Definition 1. (Zhao et al. 2005) A *Semistructured Probabilistic Object (SPO)}* S is a tuple $S = (T, V, P, C, \omega)$ where

- T is a relational tuple over some semistructured schema R^* over R. We refer to T as the *context* of S.
- $V=\{v_1,\ldots,v_q\} \subseteq \mathbf{V}$ is a set of random variables. We require that $V \neq \emptyset$, where V is called the set of *participating random variables.*
- $P: dom(V) \rightarrow \mathbf{P}$ is the *probability table* of S. Note that P need not be complete, but it must be *consistent* with respect to $\mathbf{P}$[5].

---

[5] Consistency criteria are probability-space dependent. For $\mathbf{P}_{point}$, the consistency criterion is that the sum of all probability values is less than or equal to 1. For $\mathbf{P}_{int}$ the consistency criterion is essentially equivalent to a requirement that the sum of lower bounds of each probability interval is less than or equal to 1.

- $C = \{(u_1,X_1),\ldots,(u_s,X_s)\}$, where $U = \{u_1,\ldots,u_s\} \subset \boldsymbol{V}$ and $X_i \subseteq dom(u_i)$, $1 \leq i \leq n$, such that $V \cap U = \varnothing$. We refer to C as the *conditional* of S.
- ω, called a *path,* is an expression of the probabilistic query algebra over $\boldsymbol{P}$. We define two different query algebras below.

An explanation of this definition is in order. For the SPO data model to possess the ability to store all the probability distributions described in Tables **1**, **2,** and **3 a--c**, the following information needs to be stored in a single object.

1. **Participating random variables.** These variables determine the probability distribution described in an SPO.
2. **Probability Table.** If only one random variable participates, it is a simple probability distribution table; otherwise the distribution will be joint. Probability table may be complete, when the information about the probability of every instance is supplied, or incomplete. In either case, it must be consistent, i.e., truly represent a probability distribution. It is convenient to visualize the probability table P as a table of rows of the form $(\boldsymbol{x}, \alpha)$, where $\boldsymbol{x} \in dom(V)$ and $\alpha = Prob(\boldsymbol{x}) \in \boldsymbol{P}$. Thus, we speak about rows and columns of the probability table when that makes explanations more convenient.
3. **Conditional.** A probability table may represent a distribution, conditioned by some prior information. The conditional part of its SPO stores the prior information in one of two forms: "`random variable` *u* `has value` *x*" or "`the value of random variable` *u* `is restricted to a subset` *X* `of its values`". In our definition, this is represented as a pair (u,X). When X is a singleton set, we get the first type of the condition.
4. **Context** provides supporting information for a probability distribution, information about the known values of certain parameters, which *are not considered to be random variables* by the application.
5. **Path.** Participating variables, probability table, conditional and context combined form the content of an SPO. Path, the fifth component, documents the object's history in the database in which it is stored. Objects inserted into the database receive unique object identifiers (OIDs) upon insertion. When a new SPO is constructed out of one or more existing SPOs as a result of a query algebra expression, the path of the new object will contain that expression.

2.2 The SP-algebra for Point Probabilities

Let us fix the universe of random variables $\boldsymbol{V}$, the universe of context attributes $\boldsymbol{R}$ and set the probability space $\boldsymbol{P}=\boldsymbol{P}_{\boldsymbol{point}}= [0,1]$. A finite collection $SP = \{ S_1,\ldots ,S_n\}$ of semistructured probabilistic objects over $\boldsymbol{V}$, $\boldsymbol{R}$ and $\boldsymbol{P}$ is called a *semistructured probabilistic relation (SP-relation)*. A finite collection $D = \{SP_1,\ldots,SP_r\}$ is called a *semistructured probabilistic database (SP-database).*

One important difference between semistructured probabilistic databases and traditional relational or relational probabilistic databases is that each table in a relational database has a specified schema, whereas all SP-relations are "schema-less": any collection of SPOs can form an SP-relation. This means that the division of a semistructured probabilistic database into relations is a matter of the logic of a particular application. For example, if the SP-database is built from the information supplied by three different experts, this information can be arranged into three semistructured probabilistic relations according to the origin of each object inserted in the database. Alternatively, the information can be arranged in SP-relations by the date it was obtained.

Manipulation of SPOs stored in SP-databases is done by the means of a query algebra, called the semistructured probabilistic algebra (SP-algebra). The SP-algebra contains three standard set operations: *union*, *intersection* and *difference;* it extends the definitions of standard relational operations *selection*, *projection*, *Cartesian product*, and *join* to account for the appropriate management and maintenance of probabilistic information within SPOs; in addition, it contains a new operation, *conditionalization*. The latter operation is specific to the probabilistic databases and results in the construction of SPOs that represent conditional probability distributions of the input SPOs.

Before proceeding with the description of individual operations, we define the *equality* and *equivalence* of SPOs. Two SPOs *S* and *S'* are *equal* if all their components are equal. Two SPOs are *equivalent* if their set of participating random variables, probability table, context and conditional are the same. Notice that in the case of equivalence, *paths* of two SPOs may be different. More formally,

Definition 2. (Zhao et al. 2005) Let $S = (T, V, P, C, \omega)$ and $S' = (T', V', P', C', \omega')$ be two SPOs. *S is equivalent to S'*, denoted $S \equiv S$, iff $T = T$, $V = V'$, $P = P'$ and $C = C'$.

Set Operations. Semistructured Probabilistic relations are sets of SPOs. Therefore, the definitions of union, intersection and difference of SP-relations are straightforward.

Definition 3. (Zhao et al. 2005) Let *SP* and *SP'* be two SP-relations.

- **Union:** $SP \cup SP' = \{ S \mid S \in SP \text{ or } S \in SP' \}$.
- **Intersection:** $SP \cap SP' = \{ S \mid S \in SP \text{ and } S \in SP' \}$.
- **Difference:** $SP - SP' = \{ S \mid S \in SP \text{ and } S \notin SP' \}$.

We note two features of the set operations in the SP-algebra. Classical relational algebra has a restriction on the applicability of the set operations: they are defined only on pairs of relations with matching schemas. Because SP-relations are schema-less and represent logical rather than syntactic groupings of probability distributions in an SP-database, set operations are applicable to any pair of SP-relations.

Selection. Given an SPO $S = (T, V, P, C, \omega)$, a selection query may be issued to any of its components except the path. Each part requires its own language of selection conditions. Selection on *context*, *participating random variables* and *conditionals*, when applied to an SPO, result in the SPO being selected or not in its entirety, as is the case with selection in relational algebra. Selection on *probability table* on the other hand, transforms the SPO by including in the selected object only the probability table rows that match the selection condition. For any selection operation, the path expression of the result is updated to include the selection operation. We illustrate different types of selections in the following example.

Example 1. Consider the military personnel monitoring application described in the example above. Suppose that the application database stores multiple probability distributions to be used for decision support. A human analyst working with the system may, at different times, want to see and/or use the results of the following information requests.

- `"Find all probability distributions for members of the Marine Corps."`} This is an example of selection based on context.
- `"Find all probability distributions that involve body temperature and oxygen level observations."` Body temperature and oxygen level are two of the random variables in the application domain. This is an example of selection on participating random variable.
- `"Find all probability distributions for servicemen with low oxygen levels"`. Here, the analyst wants to find what is known about the probabilities of *other* random variables in the domain, when the oxygen level (a random variable in the domain) is known to be low. This is selection on conditional.
- `"What information is available about the probability of having low oxygen level and high body temperature?"` In each SPO which contains Temperature and Oxygen variables, we are interested in the row(s)[6] of the probability table which has/have values Temperature = high and Oxygen = low. This is an example of selection on probability table.
- `"What outcomes have probability over 0.4?"` This is an example of selection on probabilities. This information need should result in only the SPOs that have probability table rows with probability values of above 0.4 returned, and *only those rows* should be shown to the analyst.

Selection on Context, Participating Variables or Conditionals. We first define the three selection operations that do not alter the content of the selected objects. We start by defining the acceptable languages for selection conditions for the three

[6] If other random variables are also present in the SPO in question, there will be more than one row matching this condition.

types of selects. Recall that the universe $\boldsymbol{R}$ of context attributes consists of a finite set of attributes $A_1,\ldots, A_n$ with domains $dom(A_1),\ldots ,dom(A_n)$. With each attribute $A \in \boldsymbol{R}$ we associate a set *Predicates(A)* of allowed predicates. We assume that *equality* and *inequality* are allowed for all $A \in \boldsymbol{R}$.

Definition 4. (Zhao et al. 2005) An *atomic context selection condition* is an expression c of the form $A \bullet x$ (or $\bullet(A,x)$), where $A \in \boldsymbol{R}$, $x \in dom(A)$ and $\bullet \in Predicates(A)$.

An *atomic participation selection condition* is an expression c of the form $v \in V$, where $v \in \boldsymbol{V}$ is a random variable.[7]

An *atomic conditional selection condition* is one of the following expressions: $u =\{x_1,\ldots, x_h\}$ or $u \ni x$ where $u \in \boldsymbol{V}$ is a random variable and $x, x_1,\ldots,x_h \in dom(u)$.

Complex selection conditions can be formed as Boolean combinations of atomic selection conditions.

Definition 5. (Zhao et al. 2005) Let $S=(T,V,P,C,\omega)$ be an SPO and let $c = A \bullet x$ be an atomic context selection condition. Let $\omega' = \sigma_c(\omega)$ and let $S' = (T,V,P,C,\omega')$. Then $\sigma_c(S) =\{S'\}$ iff:

1. $A \in S.T$;
2. For some instance A^* of A in $S.T$, $S.T.A^* \bullet x$ is true.

Otherwise, $\sigma_c(S) = \emptyset$.

Definition 6. (Zhao et al. 2005) Let $S=(T,V,P,C,\omega)$ be an SPO and let $c = v \in V$ be an atomic participation selection condition. Let $\omega' = \sigma_c(\omega)$ and let $S' = (T,V,P,C,\omega')$. Then $\sigma_c(S) =\{S'\}$ if $v \in S.V$; otherwise $\sigma_c(S) = \emptyset$.

Definition 7. (Zhao et al. 2005) Let $S=(T,V,P,C,\omega)$ be an SPO. Let $\omega' = \sigma_c(\omega)$ and let $S' = (T,V,P,C,\omega')$.

1. Let $c = u =\{x_1,\ldots ,x_h\}$ be an atomic conditional selection condition. Then $\sigma_c(S) =\{S'\}$ if $S.C \ni (u, X)$ and $X = \{x_1,\ldots,x_h\}$; otherwise $\sigma_c(S) = \emptyset$.
2. Let $c = u \ni x$ be an atomic conditional selection condition. Then $\sigma_c(S) =\{S'\}$ if $S.C \ni (u, X)$ and $x \in X$; otherwise $\sigma_c(S) = \emptyset$.

The semantics of atomic selection conditions can be extended to their Boolean combinations in a straightforward manner.

$$\sigma_{c \wedge c'}(S) ::= \sigma_c(\sigma_{c'}(S));$$
$$\sigma_{c \vee c'}(S) ::= \sigma_c(S) \cup \sigma_{c'}(S),$$

except for the path component, which will become, respectively, $\sigma_{c \wedge c'}(S)$ ($\sigma_{c \vee c'}(S)$).

---

[7] Note that "$\in V$" is syntactic sugar here. Instances of such conditions have the form $\mathit{Oxygen} \in V$, $\mathit{Condition} \in V$ and so on.

The interpretation of *negation* in the context selection condition requires some additional explanation. In order for a selection condition of the form $\neg(A \bullet x)$ to succeed on some SPO $S=(T,V,P,C,\omega)$, attribute A must be present in $S.T$. If A is not in $S.T$, then $\sigma_{\neg(A \bullet x)}(S) = \emptyset$. Therefore, the statement $S \in \sigma_c(S) \vee S \in \sigma_{\neg c}(S)$ is not necessarily true. This also applies to conditional selection conditions.

Selection on Probability Table. The two remaining types of selection operations are more complex than the three described above. Here, the result of each operation applied to an SPO can be *a non-empty part* of the original SPO. In particular, these operations preserve the context, participating random variables and conditionals in an SPO, but may return *only a subset* of the rows of the probability table. In these operations, the selection condition will indicate which rows from the probability table are to be included and which are to be omitted. In a sense, these operations treat the probability table of an SPO as a relational table, and perform selections from it.

Definition 8. (Zhao et al. 2005) An *atomic probability table selection condition* is an expression of the form $v = x$ where $v \in \boldsymbol{V}$ and $x \in dom(v)$. *Probability table selection conditions* are Boolean combinations of atomic probability table selection conditions.

Definition 9. (Zhao et al. 2005) Let $S = (T,V,P,C,\omega)$ be an SPO, $V = \{ v_1,\ldots,v_k\}$ and let $c = v = x$ be an atomic probabilistic table selection condition. Let $\omega' = \sigma_c(\omega)$. If $v \in V$, then (assuming $v = v_i$ for some $1 \le i \le k$) the result of selection from S on c, $\sigma_c(S)$, is a semistructured probabilistic object $S' = (T,V,P',C,\omega')$, where

$$P'(y_1,\ldots,y_i,\ldots,y_k) = \begin{cases} P(y_1,\ldots,y_i,\ldots,y_k) & y_i = x; \\ \textit{undefined} & y_i \neq x. \end{cases}$$

Definition 10. An *atomic probabilistic selection condition* is an expression of the form $P \bullet \alpha$, where $\alpha \in [0,1]$ and $\bullet \in \{=, \neq, \le, \ge, <, >\}$. *Probabilistic selection conditions* are Boolean combinations of atomic probabilistic selection conditions.

Definition 11. Let $S=(T,V,P,C,\omega)$ be an SPO and let $c = P \bullet \alpha$ be an atomic probabilistic selection condition. Let $\boldsymbol{x} \in dom(V)$. The result of selection from S on c is defined as follows: $\sigma_c(S) = (T,V,P',C,\omega')$, where $\omega' = \sigma_c(\omega)$ and

$$P'(\bar{x}) = \begin{cases} P(\bar{x}) & P(\bar{x}) \bullet \alpha; \\ \textit{undefined} & \neg(P(\bar{x}) \bullet \alpha). \end{cases}$$

Different selection operations commute, as shown in the following theorem.

Theorem 1. (Zhao et al. 2005) Let c and c' be two (arbitrary) selection conditions and let SP be a semistructured probabilistic relation. Then $\sigma_c(\sigma_{c'}(SP)) \equiv \sigma_{c'}(\sigma_c(SP))$.

Projection. SPOs are complex objects consisting of four different components. Traditionally, *projection* in relational algebra is a simplification operation that removes attributes. With SPOs, there are three types of simplifications that can be performed: removal of context, removal of conditionals and removal of participating random variables. All three projection operations are introduced below.

Definition 12. (Zhao et al. 2005) Let $S = (T, V, P, C, \omega)$ be an SPO and let $L \subseteq \boldsymbol{R}$ be a set of context attributes. The projection of S onto L, denoted $\pi_L(S,)$, is an SPO $S' = (T', V, P, C, \omega')$, where $T' = \{(A,x) \mid (A,x) \in T,\ A \in L\}$ (i.e., T' contains all entries from T for attributes from the list L only), and $\omega' = \pi_L(\omega)$.

Definition 13. (Zhao et al. 2005) **Let** $S = (T, V, P, C, \omega)$ be an SPO and let $F \subseteq \boldsymbol{V}$ be a set of random variables. The projection of the conditional part of S onto F, denoted $\pi_{c:F}(S)$, is an SPO $S' = (T, V, P, C', \omega')$ where $C' = \{(u,X) \mid (u,X) \in T.C,\ u \in F\}$ and $\omega' = \pi_{c:F}(\omega)$.

We note that since both the context and the conditional part of an SPO can be empty, projections $\pi_{\varnothing}(S)$ (i.e., removal of all context information for an SPO) and $\pi_{c:\varnothing}(S)$ (clearing of the list of conditionals) are valid and will yield proper results. A somewhat more complicated and delicate operation is the projection on the set of participating random variables. A removal of a random variable from the SPO's participant set entails that all information related to this random variable has to be removed from the probability table as well. This essentially corresponds to removal of a random variable from consideration in a joint probability distribution, which is usually called *marginalization*. The result of this operation is a new marginal probability distribution that needs to be stored in the probability table component of the resulting SPO.

This computation is performed in two steps. First, the columns for random variables that are to be projected out are removed from the probability table. In the remainder of the table, there can now exist duplicate rows whose values for all the fields except the probability coincide. All duplicate rows of the same type are then collapsed (coalesced) into one, with the new probability value computed as the sum of the values in the constituent rows. The formal definition of this procedure is given below.

Definition 14. (Zhao et al. 2005) Let $S = (T, V, P, C, \omega)$ be an SPO, $V = \{v_1, \ldots, v_q\}$, $\boldsymbol{q>=1}$, and let $L \subseteq \boldsymbol{R}$ be a *non-empty* set of random variables. If $L \cap S.V = \varnothing$, then the projection of S on L, denoted $\pi_L(S)$, is an empty set. If If $L \cap S.V \neq \varnothing$, then $\pi_L(S) = \{S'\}$ where $S' = (T, L, P', C, \omega')$ and where $P': dom(L) \rightarrow [0,1]$ and for each $\bar{x} \in dom(L)$,

$$P'(\bar{x}) = \sum_{\bar{y} \in dom(V-L);\, P(\bar{x},\bar{y})\ \text{is defined}} P(\bar{x},\bar{y}) \; .$$

Notice that projection on the participating random variables is allowed only if the $S.V$ is not a singleton and if at least one random variable remains in the resulting set.

Conditionalization. Conditionalization is an operation specific to probabilistic algebras. Dey and Sarkar (Dey and Sarkar 1996) were the first to consider this operation in the context of probabilistic databases. Similarly to the variable projection operation, conditionalization reduces the probability distribution table. The difference is that the result of conditionalization is a *conditional probability distribution.* Given a joint probability distribution, conditionalization answers the general query of the form, "`What is the probability distribution of the remaining random variables if the value of some random variable` *v* `in the distribution is restricted to subset` *X* `of its values?`"

Informally, the conditionalization operation proceeds on a given SPO as follows.

The input to the operation is one participating random variable of the SPO, *v*, and a subset of its domain $X \subseteq dom(v)$. The first step of the operation consists of removal from the probability table of the SPO all rows whose *v* values are not from the set *X*. Then the *v* column is removed from the table. The remaining rows are coalesced (if needed) in the same manner as in the projection operation and afterwards, the probability values are normalized. Finally, *(v,X)* is added to the set of conditionals of the resulting SPO.

The formal definition of conditionalization is given below. Note that if the original table is incomplete, there is no meaningful way to normalize the probability distribution. The operation can still be performed, but the results may be meaningless. Thus, we restrict this operation to situations where normalization is well defined.

Definition 15. (Zhao et al. 2005) An SPO $S=(T, V,P,C,\omega)$ is *conditionalization-compatible* with an atomic conditional selection condition $v=\{x_1,\dots,x_h\}$ iff (a) $v \in S.V$ and (b) the restriction of $S.P$ on $\{x_1,\dots,x_h\}$ for variable v is a complete function.

Definition 16. (Zhao et al. 2005) Let SPO $S=(T, V,P,C,\omega)$ be an SPO which is conditionalization-compatible with an atomic conditional selection condition $c = v=\{x_1,\dots,x_h\}$. The result of *conditionalization* of *S* by *c*, denoted $\mu_c(S)$, is defined as follows: $\mu_c(S) = (T,V',P',C',\omega)$, where

- $V' = V - \{v\}$;
- $C' = C \cup \{(v,\{x_1,\dots,x_h\})\}$;
- $P':V' \to [0,1]$ is defined as follows.

Let

$$N = \sum_{\bar{y} \in dom(V')} \sum_{x \in \{x_i,\dots,x_h\}} P(x,\bar{y}).$$

Then, for any $\bar{y} \in dom(V')$, P' is defined as follows:

$$P'(\bar{y}) = \frac{\sum_{x \in \{x_1,\dots,x_h\}} P(x,\bar{y})}{N};$$

- $\omega' = \mu_c(\omega)$.

Cartesian Product and Join. Sometimes an SP-database has only simple probability distributions for some random variables. In order to get a joint probability distribution, either a Cartesian product or a join operation can to be performed on the SPOs storing these distributions. Intuitively, both a Cartesian product and a join of two probabilistic distributions compute the joint probability distribution of random variables involved in both original distributions. The difference between them lies in the operation applicability. The Cartesian product can be computed only for a pair of SPOs with disjoint participating random variables. The join operation is applicable to two SPOs that share common participating random variables.

When a joint probability distribution is computed from individual (marginal) probability distributions, knowledge of the relationship between the random variables in the two marginal distributions is necessary to correctly compute the joint probability distribution. In this narrative, we restrict ourselves to the case when random variables from the two distributions are conditionally independent. This restriction allows us to represent the result as a joint probability distribution which can be explicitly computed: the joint probability is the product the marginal probabilities˙ Other assumptions that allow for direct computation of joint probability distributions are discussed elsewhere (Zhao et al. 2006).

Two SPOs are *compatible* for Cartesian product if their participating variables are disjoint, but their conditionals coincide.

Definition 17. (Zhao et al. 2005) Two SPOs $S = (T,V,P,C,\omega)$ and $S' = (T',V',P',C',\omega')$ are *Cartesian product-compatible (cp-compatible)* if and only if (a) $V \cap V' = \varnothing$ and (b) $C = C'$.

We can now define the Cartesian product.

Definition 18. (Zhao et al. 2005) Let $S = (T,V,P,C,\omega)$ and $S' = (T',V',P',C',\omega')$ be two cp-compatible SPOs. The result of their Cartesian product (under assumption of independence), denoted $S \times S'$, is: $S \times S' = S'' = (T'',V'',P'',C'',\omega'')$, where

- $T'' = T \cup T'$;
- $V'' = V \cup V'$;
- P'': $dom(V'') \to [0,1]$ is defined as follows. For all $\bar{z} \in dom(V'')$, where $\bar{z} = (\bar{x}, \bar{y})$, $\bar{x} \in dom(V)$, $\bar{y} \in dom(V')$: $P''(\bar{z}) = P(\bar{x}) \cdot P(\bar{y})$;
- $C'' = C = C'$;
- $\omega'' = \omega \times \omega'$.

The join operation extends the Cartesian product operation to the situation, where two SPOs being combined share random variables. If we have two probability distributions *Prob(X,Y)* and *Prob(Y,Z)*, then a joint probability distribution *Prob(X,Y,Z)* can be represented as *Prob(X,Y,Z)* = *Prob(X,Y)*Prob(Z|Y)* = *Prob(X|Y)* * *Prob(Y,Z)*. The two representations of the joint probability distribution (one, conditioning Z on *Y* and another, conditioning *X* on *Y*) are equal if the probability distributions are drawn from one known underlying universal probability distribution on ***V***. However, the SPO framework can store, in the same database, information from multiple universal distributions (e.g., distinguished by the context settings of the SPOs). Thus, $Prob(X,Y)*Prob(Z|Y) = Prob(X|Y)*Prob(Y,Z)$ is not necessarily always true. To make sure that SPOs can be joined efficiently, we consider two separate join operations, one using the $Prob(X,Y)*Prob(Z|Y)$ representation, and the other using the $Prob(X|Y)*Prob(Y,Z)$. These operations are known as *left join* and *right join.*

Definition 19. (Zhao et al. 2005) Two SPOs $S = (T,V,P,C,\omega)$ and $S' =(T',V',P',C',\omega')$ are *join-compatible* if and only if (a) $V \cup V' \neq \emptyset$ and (b) $C = C'$.

Given two join-compatible SPOs S and S', we can break the set $V \cup V'$ into three non-empty disjoint parts: $V_1 = V - V'$, $V_2 = V' - V$ and $V_c = V \cap V'$. The information about the probability distribution of random variables in V_c can be found in both S and S'. The join operation must take this into consideration when the joint probability distribution for variables in $V \cup V'$ is computed. The key to computing the joint distribution correctly is the following statement.

Lemma 1. Let $\mathbf{x} \in dom(V_1)$, $\mathbf{y} \in dom(V_c)$, $\mathbf{z} \in dom(V_2)$, and let V_1, Vc and V_2 all be disjoint. Under the assumption of independence between variables in V_1 and V_2 the following holds:

$$\Pr(\overline{x},\overline{y},\overline{z}) = \Pr(\overline{x},\overline{y}) * \Pr(\overline{y},\overline{z}) / \Pr(\overline{y}) = \Pr(\overline{x},\overline{y}) * P(\overline{z} \mid \overline{y}) = P(\overline{z},\overline{y}) * P(\overline{x} \mid \overline{y}).$$

We can now define the join operations. We want the join of S and S' to contain the joint probability distribution of the set $V_1 \cup V_c \cup V_2$. Since $\Pr(\mathbf{y})$ could be obtained either from S or from S', there exist two families of join operations, called *left join* and *right join*, with the following definitions.

Definition 20. (Zhao et al. 2005) Let $S = (T,V,P,C,\omega)$ and $S' =(T',V',P',C',\omega')$ be two join-compatible SPOs. Let $V = V_1 \cup V_c$ and $V' = V' \cup Vc$, and $Vc = V \cup V'$. We define the operations of *left join* of S and S', denoted $S \rhd\!< S'$, and right join of S and S', denoted $S >\!\lhd S'$, as follows:

- $S \rhd\!< S' ::= S'' = (T'',V'',P'',C'',\omega'')$;
- $S >\!\lhd S' ::= S''' = (T'', V'', P''',C'',\omega''')$, where
  1. $T'' = T \cup T'$;
  2. $V'' = V_1 \cup V_c \cup V_2$;
  3. $P''': dom(V'') \to [0,1]$ is computed as follows.

For all $\overline{w} \in dom(V'')$; $\overline{w} = (\overline{x},\overline{y},\overline{z})$; $\overline{x} \in dom(V_1)$, $\overline{y} \in dom(V_c)$, $\overline{z} \in dom(V_2)$:

$$P''(\overline{w}) = P(\overline{x},\overline{y}) \cdot \frac{P'(\overline{y},\overline{z})}{P'(\overline{y})}$$

$$P''(\overline{w}) = P'(\overline{y},\overline{z}) \cdot \frac{P(\overline{x},\overline{y})}{P(\overline{y})}$$

- $C'' = C = C'$.
- $\omega'' = \omega \rhd\!< \omega'$; $\omega'' = \omega >\!\lhd \omega'$.

Two *join-compatible* SPOs are *join-consistent* if probability distributions on the set of shared participating variables are identical for both SPOs.

Definition 21. (Zhao et al. 2005) Let Let $S = (T,V,P,C,\omega)$ and $S' =(T',V',P',C',\omega')$ be two join-compatible SPOs with $V \cap V' = V_c$. Then, S and S' are *join-consistent* if and only if $P(\overline{y}) = P'(\overline{x})$ for any $\overline{y} \in dom(V_c)$.

SP-algebra operations can be extended to a semistructured probabilistic relation, as described in the following proposition.

Proposition 1. (Zhao et al. 2005) Any SP-algebra operation on a semistructured probabilistic relation is equivalent to the union of the SP-algebra operation on each SPO in the SP-relation:

- Let SP be a semistructured probabilistic relation and γ be one of the three unary SP-algebra operators. Then $\gamma(SP) = \bigcup_{S \in SP} \gamma(S)$.
- Let SP_1 and SP_2 be two semistructured probabilistic relations and $\oplus$ be one of the binary SP-algebra operators. Then:

$$SP_1 \oplus SP_2 = \bigcup_{S \in SP_1} \bigcup_{S' \in SP_2} S \oplus S'.$$

Semantics of the SP-algebra Operations. The problem of determining the meaning of the results of the operations of the SP-algebra is complicated by the fact that at any moment, SP-databases can contain SPOs of two types. In the SPOs of the first type, the probabilities of all rows are exact, while in the SPOs of the second type, the probabilities of some rows may represent the lower bounds on the probability of those instances. We proceed by defining the two types of SPOs formally, discussing their properties and the effects that different SP-algebra operations have on the SPOs in light of this.

Definition 22. (Zhao et al. 2005) An SPO $S =(T,V,P,C,\omega)$ is a *Type I* SPO iff $\sum_{x \in dom(V)} P(\boldsymbol{x}) = 1$. Otherwise, S is a *Type II* SPO.

When S is a Type I SPO, its probability table is *complete*: the probabilities of all rows add up to exactly 1. The probability table may contain a row for every instance $\boldsymbol{x} \in dom(V)$, or it may omit some of the instances. However, because the probabilities of the rows present in the table add up to 1, we know that the probabilities of all omitted rows are 0, and these can be added to the probability table of S. Basically, when S is a Type I SPO, we are guaranteed that *for all* $\boldsymbol{x} \in dom(V)$ $P(\boldsymbol{x})$ *is the exact point probability of instance* $\boldsymbol{x}$.

The nature of Type II SPOs is somewhat more complex. If the sum of probabilities in all rows of the probability table is less than 1, then that the probability table is missing some information. This can either be *missing instances*: some $\boldsymbol{x} \in dom(V)$ has a non-zero probability but is not included in the probability table of S, or *underestimation*: all possible instances are present, but the probabilities add up to less than 1, which means that information about the probabilities of some (or all) instances is only a *lower bound* on the true probability of the instance in the distribution.

It is important to note here that SP-algebra operations allow for Type II SPOs to occur in the SP-database, even if all original SPOs in the database were Type I. The difference in the meaning of probability values for Type I and Type II SPOs causes us to apply extra caution when interpreting the results of SP-algebra operations. In particular, when considering a specific SP-algebra operation applied to an SPO or a pair of SPOs, it is important for us to know the type of the input objects and be able to determine the type of the result. The following proposition identifies the set of "safe" operations in SP-algebra: operations that, given Type I SPOs, are guaranteed to produce Type I results.

Proposition 2. (Zhao et al. 2005) Let S and S' be two Type I SPOs. Then, the following SPOs are also Type I:

1. $\sigma_c(S)$, where c is a selection condition on *context*, *participating random variables* or *conditional.*
2. $\pi_L(S)$, $\pi_{c:F}(S)$ and $\pi_F(S)$, where L is a list of context attribute names and $F \subseteq \boldsymbol{V}$.
3. $\mu_c(S)$, where c is a conditional selection condition.
4. $S \times S'$.
5. $S \rhd\!< S'$ and $S >\!\lhd S'$.

Two operations missing from the list in Proposition 4 are selection on probabilities and selection on probability table. These operations can take as input Type I SPOs and produce Type II SPOs, because both operations can return incomplete probability tables. The following statements specify the semantics of the SP-algebra operations producing Type I results.

Theorem 2. (Zhao et al. 2005) Let $S = (T, V, P, C, \omega)$ be a Type I SPO and let $\emptyset \neq L \subseteq \boldsymbol{V}$. Let $S' = (T, L, P', C, \omega') = \pi_L(S)$. Then $S'.P'$ contains the ***correct marginal probability distribution*** of random variables in L given the probability distribution $S.P$.

Theorem 3. (Zhao et al. 2005) Let $S = (T, V, P, C, \omega)$ be a Type I SPO and let c be a conditional selection condition involving variable $v \in S.V$. Let $S' = (T, V-\{v\}, P', C', \omega') = \mu_c(S)$. Then $S'.P'$ contains the ***correct conditional probability distribution*** of random variables $S.V-\{v\}$ from the distribution $S.P$ given condition c on v.

Theorem 4. (Zhao et al. 2005) Let $S = (T, V, P, C, \omega)$ and $S' = (T', V', P', C', \omega')$ be two cp-compatible SPOs and let $S'' = (T'', V'', P'', C, \omega'') = S \times S'$. Then $S''.P''$ is the correct joint probability distribution of random variables in $S.V$ and $S'.V'$ under the assumption of independence between them.

Theorem 5. (Zhao et al. 2005) Let $S = (T, V, P, C, \omega)$ and $S' = (T', V', P', C', \omega')$ be two join-compatible SPOs and let $S'' = (T'', V'', P'', C, \omega'') = S \rhd\!< S'$ and $S''' = (T''', V''', P''', C, \omega''') = S >\!\lhd S'$. Then $S''.P''$ and $S'''.P'''$ are the ***correct joint probability distributions*** of random variables in $S.V$ and $S'.V'$ under the assumption of independence between them.

Theorem 6. (Zhao et al. 2005) Let S and S' be two join-compatible SPOs. The left join S ▷< S' and the right join S >◁S'$ are equivalent if and only if S and S' are *join-consistent.*

2.3. Extensions of the SPO Framework

Interval SPO Model. As mentioned above, the probability space $\boldsymbol{P}_{point}=[0,1]$, is not the only way to represent probabilistic information in the SPO framework. Probability intervals have been, for some time, considered the *next* natural extension of the notion of probability (Walley 1991, de Campos et al. 1994, Weichselberger 2000, Ng and Subrahmanian 1992). Because the definition of an SPO factors out the probability space, a valid Semistructured Probability Object may use probability intervals rather than point probabilities in its probability table. Such SPOs were introduced in a somewhat tongue-in-cheek manner (Goldsmith et al. 2003) as the means of representing results of political surveys. While the data representation format does not change much, the same cannot be said about the semantics of the SPOs and, consequently, the query algebra. An interval probability distribution is modeled using Nilsson's (Nilsson 1986) possible worlds semantics, (Weichselberger 2000, de Campos et al. 1994): a true probability distribution assigns point probabilities to all rows in the probability table, but is unknown. Probability intervals represent a set of linear constraints on the point probabilities. An interval probability distribution will satisfy some (possibly none) point probability distributions, termed *p-interpretations* (Ng and Subrahmanian 1992, Zhao et al. 2004), each of which is considered equally likely to be the true one.

The query algebra operations were extended to preserve the mapping between interval probability distributions and the sets of satisfying p-interpretations (Zhao et al. 2004, Zhao et al. 2003). The semantics of extended (interval) SP-algebra operations that do not alter probabilities, set operations and various selection operations, does not change much from the SP-algebra case. On the other hand, projection, Cartesian product, join, and, especially, conditionalization, operations that modify probabilities, become much more involved. With the exception of conditionalization, extended SP-algebra versions of all operations preserve the possible worlds semantics: i.e., we prove that a *p-interpretation* satisfies the interval probability distribution obtained in the result of an extended SP-algebra operation if and only if it be constructed from some *p-interpretation* (or a pair of *p-interpretations*) by applying an SP-algebra analog of the operation to it/them (Zhao et al. 2004).

For the conditionalization operation, the interval distribution obtained as result of the extended operation is guaranteed to be *tight*, i.e., some *p-interpretations* satisfying the original interval probability distribution get transformed by a conditionalization operation in a way that matches all interval boundaries. However, the resulting interval probability distribution can have satisfying *p-interpretations* that cannot be obtained from any p-interpretation satisfying the original (pre-conditionalization) interval probability. This is an instance of a general result, due to Jaffray (1992), concerning computing conditional imprecise probabilities. It represents an essential structural shortcoming of the interval probability models in general, and the extended SPO framework in particular.

SPDBMS. The SPO framework was implemented by Zhao (Zhao et al. 2005) using transformation of SP-relations in collections of relational tables on top of a relational DBMS. The Semistructured Probabilistic DBMS (SPDBMS for short) supports basic data manipulation (insert, delete, update an SPO) and provides full support for the SP-algebra. Because collections of SPOs are inherently semistructured, the translation of SPOs into relational tables is rather cumbersome. More recently, SPDBMS was re-implemented using the native XML DBMS eXist (Rosson 2008). This avoided the data translation step. Query algebra operations were implemented using XQuery, and XQuery's user-defined functions. Most of the operations behaved efficiently. However, due to the specifics of eXist's internal architecture,[8] processing Cartesian products and joins was unreasonably slow.

SPOQL. The SP-algebra provides a functional query language for querying SP-databases. Direct SP-algebra syntax was implemented in both versions of SPDBMS and used as the query language. In addition to the SP-algebra, a declarative query language for SP-databases, called SPOQL, was introduced and implemented as part of the RDBMS-based SPDBMS (Dekhtyar et al. 2006).

3 Modeling Uncertain Data

In this section, we consider databases that treat *data* as uncertain, rather than storing and managing uncertainty in terms of probability distributions. We model data uncertainty in three ways: (1) tuple uncertainty, (2) attribute uncertainty, and (3) sub-attribute uncertainty.

In *tuple uncertainty*, a probability number (sometimes called confidence) is associated with each tuple. An example is shown in Figure 5(a), which is similar to Figure 1, except that we have *mutual exclusion* correlations among tuples. Recall that it is from an application in which various sensors are embedded in the uniforms of soldiers in a battle field. The sensors send out detections of the medical conditions of the soldier that wears the uniform. The second to last column is a score that indicates how much medical attention this soldier needs. The higher the score, the more urgent it is to send medical resources to this soldier. The last column (Conf.) is the probability that the tuple exists in the table.

We may specify mutual exclusion rules, which indicate that at most one of a set of tuples can exist in the table. In this way, we can encode a discrete PMF (probability mass function) by a set of mutually exclusive tuples. In more detail, for a PMF $\{(v_1, p_1), (v_2, p_2), \ldots, (v_k, p_k)\}$, v_1 to v_k are values in a set of mutually exclusive tuples and p_1 to p_k are their probabilities. The sum of the probabilities is no more than 1. If the sum is less than 1, then with remaining probability, none of the mutually exclusive tuples exist in the table. In the example in Figure 4(a), the three highlighted tuples (T2, T4, and T7) are mutually exclusive. They are detections of the same soldier (same Soldier ID) at around the same time, and hence at

[8] The version of eXist used by Rosson (Rosson 2008) loaded user-defined functions and reinterpreted them each time they were invoked, which affected the join and Cartesian product operations .

Tuple ID	Soldier ID	Time	Location (x, y)	Score for Medical Needs	Conf.
T1	1	10:50	(10, 20)	49	0.4
T2	2	10:49	(10, 19)	60	0.4
T3	3	10:51	(9, 25)	110	0.4
T4	2	10:50	(10, 19)	80	0.3
T5	4	10:49	(12, 7)	56	1.0
T6	3	10:50	(9, 25)	58	0.5
T7	2	10:50	(11, 19)	125	0.3

(a)

Tuple Uncertainty

Object ID	Speed
28	Normal (90, 20)
11	Normal (62, 15)
72	Normal (78, 10)

(b) **distribution**

Attribute Uncertainty

Fig. 5 Illustrating two kinds of uncertain data: tuple uncertainty (a) and attribute uncertainty (b). The last column of (a) (Conf., i.e., confidence) indicates the probability that the tuple exists in the table. The highlighted tuples are mutually exclusive (i.e., at most one of them can be true).

most one of them can have the correct score. The tuple uncertainty model can be considered as a generalization of the data model without uncertainty, in which each tuple has probability one, and there are no mutual exclusion rules.

The second type of uncertainty is called *attribute uncertainty*. In this case, an attribute is uncertain and we model each value of the attribute as a probabilistic distribution. In the example of Figure 5(b), the measurements of the *Speed* attribute can have errors and we model each speed value by a normal distribution. This is in contrast with the traditional deterministic model in which each value of an attribute is a fixed scalar value. Attribute uncertainty may also be considered as a generalization of the data model without uncertainty, in which each value in an attribute is some value with probability one (i.e., a discrete distribution).

Not only do the two kinds of uncertainty exist in the source data, but they also exist in the *query result*. Let us look at an example. We take a simple table that has attribute uncertainty as shown in Figure 5(b). We then issue a query as in Figure 6(a). What would the result be? Each of the three tuples has a non-zero probability to satisfy the predicate "*Speed > 78*". For example, the first tuple's Speed attribute has a normal distribution with mean 90 and variance 20, and thus has a high probability (say, 0.95) of satisfying the predicate. The second tuple, on the other hand, has a normal distribution with a low mean (62) and has a tiny probability (say, 0.001) of satisfying the predicate. Thus, we have tuple uncertainty in the query result (last column in Figure 6(a)).

What about the selected "*Speed*" attribute in the result set? We know that only if the *Speed* is above 78 should the tuple be in the result at all. Hence, we can reason that the *Speed* attribute in the result should not be in its original form, but rather, a conditional distribution (conditioned on the predicate being true) based on the original distribution. We illustrate this in Figure 6(b), which shows the example for the first result tuple. We cut off the original distribution *Normal* (90, 20) at the value 78, and only take the right side of the curve. Then, we need to normalize it (by multiplying by a constant factor) so that the function still integrates to 1, as a

SELECT *ObjectID, Speed* FROM *table*
WHERE *Speed* > 78

Result? attribute uncertainty tuple uncertainty

Object ID	Speed	Prob.
28	?	0.95
11	?	0.001
72	?	0.5

(a)

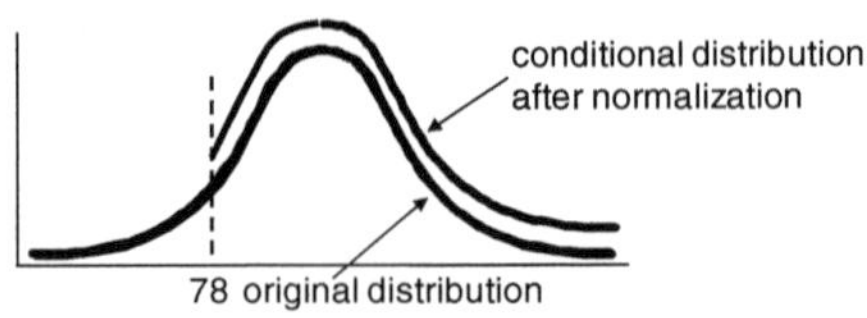

(b)

Fig. 6 Illustrating tuple uncertainty and attribute uncertainty in a query result. We issue the query in (a) to the uncertain table in Fig. 4(b). Each of the three tuples has a non-zero probability to be in the result---this is tuple uncertainty (last column in (a)). The "Speed" in the result has attribute uncertainty – a conditional distribution shown in (b).

probability density function. We can see that the *Speed* attribute in the result is still distributions, and we have attribute uncertainty in the result.

In addition, we may have *sub-attribute* uncertainty for some data types. For instance, a text string attribute can have uncertain "characters" within it. As formalized by Jestes et al. (2010), a probabilistic string can have two models: the *string-level* model and the *character-level* model, which we define next.

Definition 23 (string-level and character-level models) (Jestes et al. 2010)**.** *Let Σ be an alphabet. A probabilistic string in the string-level model is represented as $S = \{(s_1, p_1), (s_2, p_2), \dots, (s_m, p_m)\}$, where $s_i \in \Sigma^*$, $p_i \in (0,1]$, and $\sum_{i=1}^{m} p_i = 1$. A character-level probabilistic string is $S = S[1]S[2] \dots S[l]$, where each character $S[i] = \{(c_{i1}, p_{i1}), \dots, (c_{im_i}, p_{im_i})\}$, $c_{ij} \in \Sigma$, $p_{ij} \in (0,1]$, and $\sum_{j=1}^{m_i} p_{ij} = 1$. That is, a string consists of independently distributed characters, some of which can be deterministic (i.e., $p_{i1} = 1$).*

While a string-level probabilistic string follows the aforementioned attribute uncertainty (i.e., an attribute with a discrete distribution), a character-level model has distributions (of characters) embedded *inside* a string attribute, which is why it is termed *sub-attribute* uncertainty. Sub-attribute uncertainty has the finest granularity among the uncertainty models. Indeed, as shown by Ge and Li (2011), an index (for substring search) will point to uncertain character positions inside a string attribute, which can potentially be very long (e.g., millions, as in DNA strings).

4 Query Processing for Uncertain Data

We describe an algorithm that we devised to answer an *arbitrary* query on uncertain data (Ge and Zdonik 2008). The algorithm is called Statistical sampling for Equidepth Result distribution with Provable error-bounds, or SERP. SERP is essentially a Monte Carlo randomized algorithm.

4.1 The SERP Algorithm

The basic idea of a Monte Carlo algorithm for processing uncertain data is that we sample input data, run a query over the samples using a conventional query engine, and then learn a probability distribution for each random variable in the output, which includes any probabilistic field in a result tuple and a result tuple's existence probability in the result set.

The SERP algorithm uses a simple and consistent representation for both input data to queries and output query results, namely *equidepth histograms*. We consider probabilistic fields having continuous distributions. We can partition the domain of a probability density function (PDF) $f(x)$ into k intervals such that for each interval I, it holds that $\int_{x \in I} f(x)\, dx = \frac{1}{k}$. Thus, a distribution is "described" by k contiguous intervals and can be succinctly represented as $k + 1$ values indicating the boundaries of the k intervals: $(v_0, v_1, \ldots, v_k)$, where $[v_i, v_{i+1})$ is the ith interval. We assume a uniform distribution within an interval.

This is reminiscent of equidepth histograms widely used in query optimizers, and reflects the idea that the exact distribution of "high density areas" is more important and should be given higher "resolution". However, note the important difference that each bucket of an equidepth histogram contains a number of actual column values, whereas an equidepth distribution specifies the PDF of one attribute field. This representation is quite compact, only needing $k + 1$ values to describe a distribution.

Sampling from such a histogram representation is very simple: first pick one of the k intervals uniformly at random, and then pick a value from that interval uniformly at random. This sampling procedure will be used in the SERP algorithm. We consider the query execution as a black box that takes n input random variables (in general) and produce a number of output random variables. The n input random variables are either binary random variables indicating input tuple probability or probabilistic fields in the form of equidepth histograms as described above. Without loss of generality, we only need to consider how we obtain the distribution of one of the output probabilistic fields. Other fields are obtained in the same way. For example, for SUM or AVG, the inputs are the n uncertain fields in n tuples and the output is the result. The SERP algorithm (Ge and Zdonik 2008) is shown below.

Algorithm SERP($X_1, \ldots, X_n$)

Input: $X_1, \ldots, X_n$: probabilistic fields in equidepth histogram distributions
Output: the distribution of one output field

// Do the main loop of the algorithm.
// k is the number of intervals in a distribution, μ is to be determined later

for each $i \leftarrow 1, \ldots, k\mu$ **do**
    Sample each input $X_1, \ldots, X_n$ and get $x_1, \ldots, x_n$
    Run the query over $x_1, \ldots, x_n$ and let the output be y_i
end for
// Get the output distribution
Sort the output values as $y_1, \ldots, y_{k\mu}$ where $y_i \le y_{i+1}$

6: Get k contiguous intervals, each containing μ output values; the first interval contains $y_1, \ldots, y_\mu$, the second contains $y_{\mu+1}, \ldots, y_{2\mu}$, and so on. More precisely, let $v_0, \ldots, v_k$ be the interval boundaries, where $v_i = \frac{y_{i\mu}+y_{i\mu+1}}{2}$ $(1 \le i \le k-1)$, $v_0 = 2y_1 - y_2$, and $v_k = 2y_{k\mu} - y_{k\mu-1}$.
7: Return the k contiguous intervals above as the result distribution.

In the algorithm, μ is a parameter that balances accuracy with performance, as we will investigate in the analysis. Note that we model all inputs as uncertain. In reality, some input values can be certain. It is straightforward to extend the algorithm to the mixed case. Also note that from one execution on the n samples to the next, to be more efficient, we can share the query plan (i.e., the query is compiled only once, and executed many times for each loop). Further, among different executions, sub-results of parts of the query plan that only refer to data without uncertainty can be shared. Another key optimization is on I/O cost. The database engine can pay the I/O cost only once, and incrementally carry out the multiple rounds of computation in parallel. It is easy to see that SERP is scalable. The cost is no more than a constant factor of that of the same operation on data without uncertainty, regardless of the number of tuples. Additionally, SERP works even if there is correlation between different inputs. We just need to carry out the sampling from the joint distribution.

4.2 *Analysis of SERP*

We measure the distance between the result distribution computed by some algorithm and an "ideal" one based on the same input distributions, but given as much computing resources as needed. We use a well-known distance metric: variation distance.

Definition 24 (variation distance) (Mitzenmacher and Upfal 2005). *The* variation distance *between two distributions D_1 and D_2 (each being a discrete probability distribution) on a countable state space S is given by* $VD(D_1, D_2) = \frac{1}{2}\sum_{x \in S} |D_1(x) - D_2(x)|$.

We first give some insights on the variation distance metric, as we will be using it for analysis.

Lemma 2 (Mitzenmacher and Upfal 2005). *Consider two distributions D_1 and D_2. For a state x in the state space S, if $D_2(x) > D_1(x)$, then we say D_2 overflows at x (relative to D_1) by the amount $D_2(x) - D_1(x)$. Likewise, if $D_2(x) < D_1(x)$, then we say D_2 underflows at x (relative to D_1) by an amount of $D_1(x) - D_2(x)$. We denote the total amount that D_2 overflows (and underflows, respectively) as P_{over} (and P_{under}, respectively). Then, $P_{over} = P_{under} = VD(D_1, D_2)$.*

We are now ready to present a novel proof that SERP has a nice bound on the variation distance between its result distribution and the ideal one, even though we do not know the exact form of the ideal result distribution, nor do we make any assumption on how to obtain it.

Theorem 7 (Ge and Zdonik 2008). *In the SERP algorithm, let* k *and* μ *be parameters as described in the algorithm. Then, with probability at least* $1 - k \cdot \left[\left(\frac{e^{2\delta}}{(1+2\delta)^{1+2\delta}}\right)^{\mu} + \left(\frac{e^{-2\delta}}{(1-2\delta)^{1-2\delta}}\right)^{\mu}\right]$, *the variation distance between the result distribution and the ideal one is no more than* δ $(0 < \delta < 0.5)$.

Proof. Consider any one interval I of the ideal distribution. Define $k\mu$ random variables X_i $(1 \le i \le k\mu)$ as follows:

$$X_i = \begin{cases} 1, & if\ y_i \in I\ in\ line\ 3\ of\ SERP \\ 0, & if\ y_i \notin I\ in\ line\ 3\ of\ SERP \end{cases}$$

Because I is an interval of the ideal distribution, from the definition of the equi-depth partition, we have $\Pr[X_i = 1] = \frac{1}{k}$, and hence $E[X_i] = \frac{1}{k}$. We define a random variable $X = \sum_{i=1}^{k\mu} X_i$, indicating the number of output y_i's that fall in I. From the linearity of expectation, we have $E[X] = k\mu \cdot \frac{1}{k} = \mu$. As X is the sum of independent 0/1 random variables, we can apply Chernoff bounds that for any $0 < \delta < 0.5$, we have $\Pr[X > (1 + 2\delta)\mu] < \left(\frac{e^{2\delta}}{(1+2\delta)^{1+2\delta}}\right)^{\mu}$ and $\Pr[X < (1 - 2\delta)\mu] < \left(\frac{e^{-2\delta}}{(1-2\delta)^{1-2\delta}}\right)^{\mu}$. Then from the union bound,

$$\Pr[X > (1 + 2\delta)\mu\ or\ X < (1 - 2\delta)\mu] < \left(\frac{e^{2\delta}}{(1+2\delta)^{1+2\delta}}\right)^{\mu} + \left(\frac{e^{-2\delta}}{(1-2\delta)^{1-2\delta}}\right)^{\mu}.$$

Now consider all k intervals and apply the union bound again:

$$\Pr[\exists interval\ st. |X - \mu| > 2\delta\mu] < k \cdot \left[\left(\frac{e^{2\delta}}{(1+2\delta)^{1+2\delta}}\right)^{\mu} + \left(\frac{e^{-2\delta}}{(1-2\delta)^{1-2\delta}}\right)^{\mu}\right].$$

Hence,

$$\Pr[\forall interval, |X - \mu| \le 2\delta\mu] \ge 1 - k \cdot \left[\left(\frac{e^{2\delta}}{(1+2\delta)^{1+2\delta}}\right)^{\mu} + \left(\frac{e^{-2\delta}}{(1-2\delta)^{1-2\delta}}\right)^{\mu}\right].$$

Thus, with probability at least $1 - k \cdot \left[\left(\frac{e^{2\delta}}{(1+2\delta)^{1+2\delta}}\right)^{\mu} + \left(\frac{e^{-2\delta}}{(1-2\delta)^{1-2\delta}}\right)^{\mu}\right]$, all intervals contain sample result points whose number differs from the expected value by no more than $2\delta\mu$. As each such point carries weight $\frac{1}{k\mu}$ into the probability, and there are either no more than $\frac{k}{2}$ overflow intervals (holding more than μ points) or no more than $\frac{k}{2}$ underflow intervals, from Lemma 2, we get that the variation distance is no more than $2\delta\mu \cdot \frac{1}{k\mu} \cdot \frac{k}{2} = \delta$. □

To get a numerical sense about the bound, we take $k = 5$, $\delta = 0.2$, and $\mu = 60$. Then from Theorem 7, using 300 sample points (rounds), with probability at least 0.91, the variation distance between the result of the SERP algorithm and the ideal distribution is no more than 0.2. This is a (rather conservative) theoretical guarantee, and our experiments (Ge and Zdonik 2008) showed that, in practice, one can obtain a small variation distance with significantly fewer rounds. On the other hand, theoretical guarantees are important as they hold for any dataset while the result of a particular experiment depends on its data.

4.3 Join Query Semantics

We now focus on an important kind of query, namely join queries, on uncertain attributes. We show that there are two useful types of join operations specific to uncertain attributes: *value join* (v-join) and *distribution join* (d-join) (Ge 2011). V-join is a natural extension of the join operation on deterministic data. Let us first look at an example.

Table R

ID	Temperature
1	N (78, 5)
2	U (70, 75)
3	N (86, 10)
⋮	⋮

Table S

ID	Temperature
1	N (85, 6)
2	U (92, 94)
3	N (77, 8)
4	hist(...)
⋮	⋮

Fig. 7 Illustrating v-join between two uncertain attributes

Example 2 (v-join). *In Figure 7, we would like to examine the temperature attributes in table R and in table S, and find pairs that are very close. Note that both temperature attributes are uncertain and contain distributions, which appear in various forms. For instance, N(78, 5) denotes a normal distribution with mean 78 and variance 5, while U(70, 75) is a uniform distribution in the range [70, 75] and "hist(...)" indicates a histogram representation whose details we omit for clarity. The query is:*

SELECT *R.ID, S.ID* **FROM** *R, S*
WHERE $R.temperature \underset{1.0,0.8}{=} S.temperature$

This is called probabilistic threshold join query *in previous work (Cheng et al. 2006). The interpretation of the join predicate is that with probability at least* 0.8, *the difference between the two join attributes is no more than* 1.0 *degree, i.e.,* $|R.temperature - S.temperature| \leq 1.0$.

For uncertain attributes (either numerical or categorical), there is a special kind of join, which we call d-join. The idea of d-join is to treat probability distributions as "objects" and the join operation is based on the *similarity* of two distributions. We now look at some examples.

Example 3 (sensor fusion). *For high availability, five sensors redundantly measure the same environmental physical property (e.g., temperature) in a sensor network deployment on Great Duck Island (off the coast of Maine)* (Szewczyk et al.

2004). *Due to the harsh environment and the unreliable nature of the sensors, the readings can have large errors. A central database system performs a sensor fusion and uses machine learning techniques (e.g., kernel methods)* (Bishop 2007) *to obtain a temperature distribution from the five sensors. We record the temperature distributions at various times within two months in two tables (one for each month). We want to query for two time instances (one from each month) that have close temperatures.*

Example 4 (data integration). *Consider data integration from several sources. We need to perform schema matching and record linkage to combine different versions of the same data entity. However, due to schema and format inconsistencies, a data entity can have a lot of uncertainty. In the integrated database, we model the uncertainty with distributions (for either numerical or categorical values)* (Dong et al. 2009). *If two entities have similar distributions, then they are likely to be close. It is useful to find out this information.*

Example 5 (prediction queries). *We use different statistical models to predict the stock prices of a large number of companies one week from now* (Brockwell and Davis 2002). *Different models gave different results and again, by using techniques such as kernel methods* (Bishop 2007), *we can get a distribution of the predicted price of each company, which is stored in relational tables. The query is to ask for pairs of two companies that are likely to have very close stock prices at that time.*

In all these three examples, if we were to use v-join, even if two distributions are exactly the same, the probability that the join predicate is satisfied might still be *insignificant*. Here is a simple example. Suppose in Example 3, the five sensors give readings that are quite different (the difference is more than the v-join value difference parameter ε). Thus, the integrated temperature distribution has approximately five buckets, each with the same probability (1/5). Even if we were to do a v-join on two *identical* distributions as such, the probability that they are within ε apart would be only about $\sum_{i=1}^{5}\left(\frac{1}{5}\times\frac{1}{5}\right)=\frac{1}{5}$ (i.e., when both random variables fall into the same bucket). The observation here is that whether v-join is satisfied or not heavily depends on the "width" of the two distributions (i.e., the uncertainty, or, the entropy). V-join does not compare the two distributions *themselves*: two identical distributions may still fail to match. However, in all these examples, the fact that two *distributions* are close is also useful: it tends to indicate a special relationship of the two tuple entities that are being joined; i.e., their uncertain attributes are likely to be close in spite of the uncertainty. Essentially, we treat probability distributions themselves as *objects* and we are joining such objects. We are now ready to formalize v-joins and d-joins.

Definition 25 (domain partition scheme) (Ge 2011). *The* domain partition scheme *for an uncertain attribute is a many-to-one mapping of values in the domain of the uncertain attribute to a countable number of states.*

Example 6 (domain partition scheme). *If the domain of an uncertain attribute is all positive real numbers, then one possible domain partition scheme is based on a parameter* step*: we map all attribute values in the interval (0, step] to state 1, all values in (step, 2×step] to state 2, and so on.*

Consider two relations *R* and *S* that have uncertain attributes *R.A* and *S.B*. In *R*, each record's *A* attribute is a probability distribution, rather than a single value, as in deterministic databases. The distribution can be encoded in various ways, including well-known distributions (e.g., a normal distribution) and histograms. The same is true for *S.B*.

We denote a join operation between *R* and *S* on attributes *R.A* and *S.B* as $R \bowtie_{A\theta B,\varepsilon,p} S$, where ε and p are optional parameters. There are two types of join: *value join* (*v-join*) and *distribution join* (*d-join*). A v-join has a join predicate that is an (approximate) equality or an inequality with some probability threshold. For example, a v-join predicate can be $R.A \underset{\varepsilon,p}{=} S.B$, which means $\Pr(|R.A - S.B| < \varepsilon) \geq p$. This is a probabilistic version of a *band join* (DeWitt et al. 1991); for deterministic data, when the predicate is $|R.A - S.B| < \varepsilon$, it is a band-join. Another example is $R.A \underset{p}{<} S.B$, which means $\Pr(R.A < S.B) \geq p$ (ε is not present here). ε usually denotes a small value and p is a probability threshold. Note that, when it is clear from the context, we often use *R.A* to denote the random variable that represents the *A* attribute of *a tuple* in *R*, and likewise for *S.B*. A d-join predicate is denoted as $R.A \underset{\varepsilon}{\sim} S.B$. It is equivalent to $VD(R.A, S.B) \leq \varepsilon$, where $VD(R.A, S.B)$ denotes the *variation distance* (Definition 24) between a distribution in *R.A* and a distribution in *S.B*, and *R.A* and *S.B* have a common set of states resulting from their domain partition schemes (Definition 25).

4.4 *Efficiently Processing V-joins*

4.4.1 Using the First Two Moments

Our query processing techniques for v-join are based on probability theory. Specifically, the *k*th *moment* of a random variable *X* is defined as $E[X^k]$. The moments are a concise way to describe the nature of the distribution of a random variable. The first moment is the expectation of the random variable while the first two moments determine the variance of the random variable: $\mathrm{Var}[X] = E[X^2] - (E[X])^2$. In fact, all moments of a variable together uniquely define its distribution (Mitzenmacher and Upfal 2005). Simply computing and storing the first two moments (or equivalently, the *expectation* and *variance*) of a random variable (in our context, an *uncertain attribute of a record* is a *random variable*) incurs little overhead but, as we show, is very useful in making quick decisions during v-join in order to improve the speed. In some well-known distributions, such as Gaussian, the expectation and variance come for free, since they are part of the description of the distribution.

Probabilistic Band Join. Perhaps the most often used v-join is the probabilistic band join; i.e., when the join predicate is $R.A \underset{\varepsilon,p}{=} S.B$. The basic method for evaluating this predicate is by computing a double integral of the form $p' = \int_{-\infty}^{\infty} f_1(x) \int_{x-\varepsilon}^{x+\varepsilon} f_2(y) dy\, dx$, where x is a random variable in *R.A*, y is a random variable in *S.B* and $f_1(x)$ and $f_2(y)$ are the density functions of x and y, respectively. The result p' is the probability that *R.A* and *S.B* (of two tuples) are at most ε apart. Note that the v-join predicate is satisfied if and only if $p' \geq p$.

The problem with the above solution is that it is very CPU expensive. Therefore, we wish to use probability bounds to improve the speed of evaluating such a predicate. Define a random variable $X = R.A - S.B$. Then the predicate is equivalent to:

$$\Pr[|x| < \varepsilon] > p. \tag{1}$$

Let $E(X) = E(R.A) - E(S.B) = \lambda$. Then we have the following three cases, as shown in Figure 8.

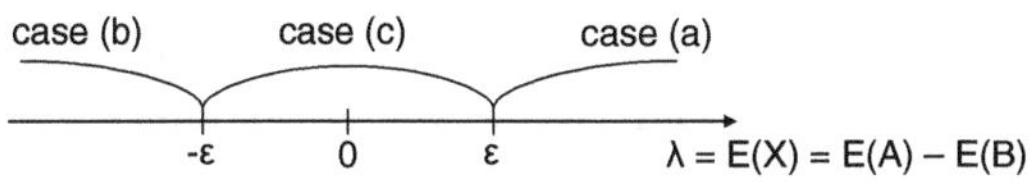

Fig. 8 Illustrating the three cases of λ

Case (a): $\lambda > \varepsilon$. We would like to know if (1) must be false, in which case we can exclude the tuple pair. From the *Cantelli's inequality* (Grimmett and Stirzaker 2001), we have:

$$\Pr[|X| < \varepsilon] \leq \Pr[X < \varepsilon] = \Pr[\lambda - X > \lambda - \varepsilon] < \frac{Var(X)}{Var(X)+(\lambda-\varepsilon)^2}.$$

If $\frac{Var(X)}{Var(X)+(\lambda-\varepsilon)^2} \leq p$, then condition (1) must be false.

Case (b): $\lambda < -\varepsilon$. Similar to case (a), we would like to use Cantelli's inequality to see if we can determine that (1) must be false and rule out the tuple pair:

$$\Pr[|X| < \varepsilon] \leq \Pr[X > -\varepsilon] = \Pr[X - \lambda > -\varepsilon - \lambda] < \frac{Var(X)}{Var(X)+(\lambda+\varepsilon)^2}.$$

If $\frac{Var(X)}{Var(X)+(\lambda+\varepsilon)^2} \leq p$, then condition (1) must be false.

Case (c): $-\varepsilon < \lambda < \varepsilon$. In contrast to the previous two cases, we would like to see if (1) must be *true* and hence the tuple pair satisfies the v-join condition. From Cantelli's inequality, we have:

$$\Pr[X > \varepsilon] = \Pr[X - \lambda > \varepsilon - \lambda] < \frac{Var(X)}{Var(X)+(\varepsilon-\lambda)^2},$$

$$\Pr[X < -\varepsilon] = \Pr[\lambda - X > \varepsilon + \lambda] < \frac{Var(X)}{Var(X)+(\varepsilon+\lambda)^2},$$

$$\Pr[|X| < \varepsilon] = 1 - \Pr[X > \varepsilon \text{ or } X < -\varepsilon] > 1 - \frac{Var(X)}{Var(X)+(\varepsilon-\lambda)^2} - \frac{Var(X)}{Var(X)+(\varepsilon+\lambda)^2}.$$

The last inequality is due to the union bound. Thus, if $1-\frac{Var(X)}{Var(X)+(\varepsilon-\lambda)^2}-\frac{Var(X)}{Var(X)+(\varepsilon+\lambda)^2}\geq p$, then condition (1) must be true.

Now suppose we are *only* given the moments of the two fields being joined. For clarity, we write *A* for *R.A* and *B* for *S.B*. For the above methods to work, we need to express λ and Var(*X*) using the moments of *A* and *B*. From the linearity of expectation, λ = E(*X*) = E(*A*) – E(*B*). With the typical assumption that *A* and *B* are independent, we have Var(*X*) = Var(*A*) + Var(*B*) = $E(A^2) - E^2(A) + E(B^2) - E^2(B)$.

Therefore, only using the first two moments of *A* and *B*, we can quickly exclude the (*A*, *B*) pair from the join result (Cases a and b) or include it in the result (Case c) if the conditions in those cases are met. If the pair is neither excluded nor included, we need to resort to the "old-fashioned" way of computing the actual probability that $|A - B| < \varepsilon$ by a double integral (or summation if they are discrete) as in (1) to see if it is greater than *p*. We can save a great deal of computational cost by using moments and probabilistic bounds to make quick judgments.

Other Inequality V-joins. Thus far we have only considered probabilistic band join; we now turn to other inequality v-joins. We only demonstrate $R.A \underset{p}{<} S.B$; we can apply similar techniques to other inequalities. Again we define a random variable *X* = *R.A* – *S.B*. Let E(*X*) = λ. We now examine two cases:

Case (a): λ < 0. Then,

$$\Pr[R.A \geq S.B] = \Pr[X \geq 0] = \Pr[X - \lambda \geq -\lambda] < \frac{Var(X)}{Var(X)+\lambda^2}.$$

If $\frac{Var(X)}{Var(X)+\lambda^2} \leq 1-p$, it must be true that $\Pr[R.A < S.B] \geq p$ and the predicate is satisfied.

Case (b): $\lambda \geq$ 0. Then we see if we can exclude the tuple pair:

$$\Pr[R.A < S.B] = \Pr[X < 0] = \Pr[\lambda - X > \lambda] < \frac{Var(X)}{Var(X)+\lambda^2}.$$

If $\frac{Var(X)}{Var(X)+\lambda^2} \leq p$, it must be true that $\Pr[R.A < S.B] < p$ and the tuple pair is excluded from the result. Details such as obtaining λ and Var(*X*) from the moments of *R.A* and *S.B* are the same as in the discussions for probabilistic band join.

We also devise indexing techniques for v-join queries. For additional details, we refer the reader to Ge (2011).

4.5 *Efficiently Processing D-joins*

4.5.1 The Condensed D-join Algorithm

In this section, we examine how we can process d-join queries efficiently. We can perform a d-join on two uncertain attributes if their domain partition schemes (Definition 3) result in a common set of states. Let the size of the state space *S* resulting from the domain partition schemes be n. Then the "features" of an uncertain distribution with respect to S can be described as $(p_1, p_2, \ldots, p_n)$, meaning that the uncertain field has probability p_i of being in state s_i. By taking this vector, we can map an uncertain distribution to a point in the n-dimensional space. It is then easy to verify that the variation distance between two distributions exactly maps to half of the L1 distance between the two corresponding points in the n-dimensional space.

This discussion leads us to the direction that a d-join can be reduced to a *similarity join,* which is well studied in the database literature (e.g., Koudas and Sevcik 2000). There are many competing algorithms that can do similarity join. However, there is a common phenomenon among the algorithms: due to the "curse of dimensionality": as dimensionality increases, performance deteriorates significantly. For example, as shown in Koudas and Sevcik's Figure 17 (2000), the response time grows 17fold as the dimensionality increases from 3 to 20 with the same number of data points for both algorithms, as shown by Koudas and Sevcik (2000). We therefore propose an algorithm called *condensed d-join*, as shown below. The algorithm starts by reducing the dimensionality by a procedure called a *condensation scheme,* as we now define.

Definition 26 (condensation scheme) (Ge 2011). A *condensation scheme* for an uncertain attribute is an onto function f: $S \rightarrow S'$, where S is the original state space determined by the domain partition scheme of the attribute and S' is a state space with a smaller cardinality, i.e., $|S'| < |S|$. The space S' is called the *condensed state space*.

Algorithm CONDENSED-D-JOIN $(R.A, S.B, \varepsilon)$

Input: Two uncertain attributes *R.A* and *S.B* whose domain partition schemes have the same states; and value ε.

Output: Pairs of *R.A* and *S.B* that satisfy $R.A \underset{\varepsilon}{\sim} S.B$.

1: ***Precomputation step:*** Determine the best condensation scheme for either *R.A* or *S.B* using the algorithm in Section 4.5.2.

2: In the condensed state space determined in line 1, we get the new distributions for all fields in *R.A* and *S.B*. If a new state is the merge of a number of previous states, then its probability is the sum of the probabilities of the original states.

3: ***Phase 1:*** Use any existing similarity join algorithm to compute the join result based on the smaller state space, using the L1 distance metric and the distance parameter 2ε.

4: ***Phase 2:*** Among the qualified tuple pairs selected in line 3, further refine the selections by computing the variation distance (*VD*) over the original state space.

Line 1 of the algorithm is to determine the optimal condensation scheme according to any one side of the join and is typically pre-computed. The condensation algorithm combines a number of neighboring states into one and sums up their probabilities. We thus get the new probability distributions in line 2. Phase 1 of the condensed d-join is performed in line 3, where we essentially reduce the d-join problem to the similarity join of multidimensional points (with the parameter value 2ε). Because of the reduced dimensionality, it is much faster. Over the qualified tuple pairs, we perform the second phase, which is a post-processing as shown in line 4. The goal of the two phase approach is to avoid the slow performance caused by high dimensionality. The quality of the condensation scheme is of a critical role here since it impacts the number of false positives that must be filtered out in the post-processing phase. We study the optimal condensation scheme in detail in Section 4.5.2.

We show the correctness of the CONDENSED-D-JOIN algorithm.

Lemma 3 (Ge 2011). *Over the original state space determined by the domain partition scheme, let distribution D_1 come from R.A and D_2 come from S.B. After the condensation, let the distributions (in step 2) be D_1' and D_2'. Then $VD(D_1', D_2') \leq VD(D_1, D_2)$.*

Proof. Suppose the condensation scheme merges k states $s_1, s_2, \ldots, s_k$ into a single state s'. Let D_1 have probabilities $p_{11}, p_{12}, \ldots, p_{1k}$ and D_2 have probabilities $p_{21}, p_{22}, \ldots, p_{2k}$ in those states, respectively. Then step (2) of the algorithm indicates that D_1' has probability $p_1' = p_{11} + p_{12} + \ldots + p_{1k}$ in state s' while D_2' has probability $p_2' = p_{21} + p_{22} + \ldots + p_{2k}$ in state s'. It holds that:

$$\begin{aligned} |p_1' - p_2'| &= |(p_{11} - p_{21}) + (p_{12} - p_{22}) + \ldots + (p_{1k} - p_{2k})| \\ &\leq |p_{11} - p_{21}| + |p_{12} - p_{22}| + \ldots + |p_{1k} - p_{2k}|. \end{aligned}$$

Thus, iterating this over all states of D_1' and D_2', summing up the inequalities as produced above, and finally dividing both sides of the resulting inequality by 2, we get $VD(D_1', D_2') \leq VD(D_1, D_2)$, which directly follows from the definition of VD. □

Theorem 8 (Ge 2011). *The* CONDENSED-D-JOIN *algorithm gives the correct result.*

Proof. From Lemma 3, we know that if, in the original state space, the VD between two distributions is less than ε, then it must also be true after the condensation. Thus, phase 1 of the d-join (step 3) will not miss any result tuples that should be returned. Finally, the second phase of the algorithm filters out all false positives. □

4.5.2 The Optimal Condensation Scheme

Consider an uncertain attribute and a condensation scheme that reduces the number of its states from n to k ($k < n$). The question now is how we should merge the states in the original state space S. Let us look at a motivating example. Figure 9(a) shows the distributions of an uncertain attribute. The solid vertical lines describe the domain partition scheme: an interval between two neighboring lines is a state. Suppose the (blue) dotted lines indicate a potential condensation scheme: there are three condensed states: the interval $[v_1, v_2)$ is the first condensed state, $[v_2, v_3)$ is the second, and $[v_3, v_4]$ is the third. It appears to be a fair condensation scheme as each condensed state contains about the same number of the original states. However, let us suppose that 1000 distributions fall in the middle range, i.e., between v_2 and v_3, while there is only one distribution in the first and third condensed states, respectively.

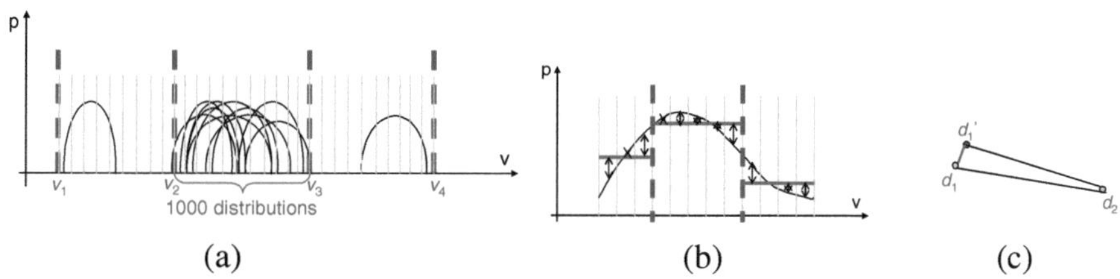

Fig. 9 (a) The necessity of a good condensation scheme, (b & c) illustrating the concept of "minimum disturbance" of a condensation scheme as measured by variation distance

Then this condensation scheme loses a lot of information: all of the 1000 distributions have the same distribution (0, 1, 0) in the condensed state space (each number is the probability of one state). In other words, it is not discriminative. In the condensed space, if one of the 1000 distributions matches with a distribution in another column for d-join, so will all other 999 distributions. We therefore need a principled algorithm to make the condensation scheme more *discriminative*.

But how to make it discriminative? The idea is to make the new distributions after applying the condensation scheme as *faithfully* as possible to the original ones. The faithfulness is again measured by variation distance. Clearly condensation would lose some information about the distributions. Thus, we would prefer a scheme that would result in new distributions that have the minimum distance from the original ones, which we call *minimum disturbance*. It is quantified by the sum of the variation distances between each new distribution and its original one, in the *original* state space. Since a new distribution can be considered as a *lossy compression* of the original one, when we convert the new distribution back to its original state space, we simply divide the probability of a condensed state by the number of the original states that map to it. This is because we do not distinguish between those states in the condensed space.

Therefore, when computing the variation distance between an original distribution and the new one, we compare the probabilities of the original states with their averages in each group, where each group corresponds to a condensed state. We illustrate this in Figure 9(b). In Figure 9(c), the original distribution is mapped to a point d_1 in the multi-dimensional space. For d-join we need to compute the distance between d_1 and a point d_2 that represents a distribution in another column. The condensation step brings d_1 to another point d_1'. The point d_1' corresponds to the conversion of the new distribution back to its original state space, which we describe in the previous paragraph. Even though d_2's position is not known *a priori*, by minimizing the distance between d_1 and d_1', the distance between d_1 and d_2 is optimally approximated by the distance between d_1' and d_2. Moreover, this optimization problem is over *all* distributions in an uncertain column.

We first formalize the problem (Ge 2011). An uncertain column has N probability distributions. The column follows a domain partition scheme that consists of n states in some serial order (e.g., n small buckets in value order). The goal of our condensation scheme is to merge some neighboring states in order to reduce them to k states ($k < n$). The scheme is chosen in such a way that the variation distance between the new distribution and the original one, summing over all records of the column, is minimized.

Let us denote the optimal (i.e., minimum) sum value (over the whole column) of the *variation distances* between the new distributions and the original ones as $D(k, n)$, where k is the target number of condensed states and n is the original number of states. We then have the following recursion:

$$D(k,n) = \min_{k \le i \le n} [\, D(k-1, i-1) + \sum_{r=1}^{N} C^{(r)}(i,n) \,] \tag{2}$$

where $C^{(r)}(i, n)$ is the "cost" of merging states from i to n into a single new state for the distribution in record r. This cost is just the part of the variation distance between the two distributions at states from i to n. More precisely (recall Figure 8 b & c),

$$C^{(r)}(i,n) = \frac{1}{2}\sum_{j=i}^{n} |\, p_j^{(r)} - a_i^{(r)} \,|, \; where \; a_i^{(r)} = \frac{1}{n-i+1}\sum_{j=i}^{n} p_j^{(r)} .$$

(3)

Here $p_j^{(r)}$ is the probability of the jth state in the rth distribution. We also note the boundary condition that

$$D(1,i) = \sum_{r=1}^{N} C^{(r)}(1,i), \quad for\; 1 \le i \le n-k+1 \tag{4}$$

We then can have an efficient dynamic programming algorithm for this problem, as illustrated in Figure 10(a). The figure shows a "D table" for values of the D function in Equation (1). The row numbers of the D table (1 to k) are the first parameter of the function while the column numbers (1 to n) are the second parameter. Our target value is $D(k, n)$, which is indicated by the red "?" at the bottom right corner of the D table. From the recursion in Equation (1), the target value can be obtained from the values in the row above, assuming we already have all the C values. The whole process can be recursively applied for each cell in the table. We therefore have a top-down procedure to fill in the shaded region in Figure 10(a) row by row, starting from the boundary condition as described in Equation (4).

In the above algorithm, we assume that we have all the C values. We now describe how to obtain them. We simply do a scan of the whole uncertain column and compute the aggregation of the C values as described in Equation (3). Figure 10(b) illustrates the C table. It is not hard to verify from Equations (2) and (4) that we only need to fill in the shaded region of the C table (C is a two-dimensional array). As we scan the column and get each distribution, we obtain the C values of the shaded region using Equation (3). Because in Equations (2) and (4) we require a sum of the C values over all distributions, we do the aggregation (sum) for each cell of the shaded region of the C table as we scan each distribution of the column

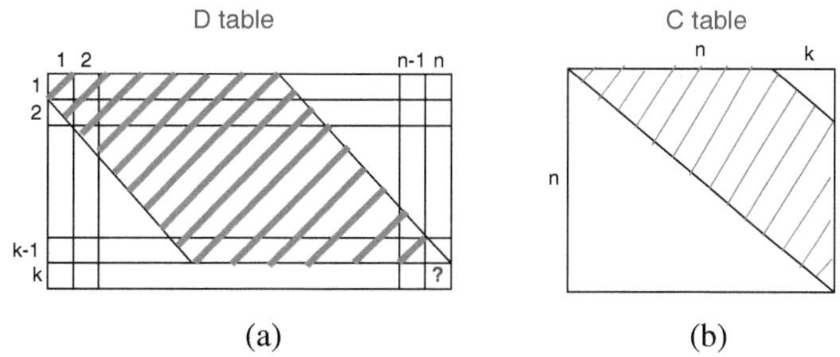

Fig. 10 A dynamic programming algorithm to get the D function (a) and a column scan to obtain the aggregated C values (b)

one by one. Eventually, when we finish scanning the uncertain column, each cell of the C table contains a sum value. Combining the above two algorithms (i.e., getting the C table followed by getting the D table), we have an efficient method to obtain the optimal condensation scheme.

5 Related Work

There has been substantial work on managing uncertain data and information in recent years due to the rise of new applications that demand this capability.

Nilsson's seminal paper on probabilistic logic (Nilsson 1986) introduced the notion of reasoning with probabilities to the field of artificial intelligence. The possible worlds semantics proposed by Nilsson has become a *de facto* standard for interpreting statements about probabilities both in the field of AI and, a bit later, in the field of databases. The key idea expressed by Nilsson is that in a world with multiple discrete random variables with finite domains, each random variable must take an exact value, and the uncertainty essentially expresses the lack of information an observer has. Each assignment of values to all random variables in the model (universe) is known as a *possible world.* If a probability is associated with each possible world, then a probability of a random variable taking a specific value is computed as the sum of probabilities of all possible worlds in which this assignment occurs.

The first work on probabilistic databases did not directly use the possible world semantics. However, the proposed frameworks were consistent with it. Cavallo and Pittarelli (1987) were perhaps the first who studied probabilistic data in the context of databases. They proposed a framework in which a probabilistic relation represented a single probability distribution. Tuples in such relations represented the probabilities of specific outcomes. Cavallo and Pittarelli defined two query algebra operations for working with such data: join, which produced a joint probability distribution for a pair of probabilistic relations, and selection, which returned the probabilities of specific outcomes.

Most of the work that followed the work of Cavllo and Pittarelli (1987), however, adopted a different view about what should be represented by a probabilistic relational table. Just as individual tuples in classical relational tables represent *independent* statements of fact, in these approaches each tuple in a probabilistic table represents a single probability distribution. The first to propose this approach in early 1990s were Barbara, Garcia-Molina and Porter (1992). In their framework a relation had a set of *certain* attributes that jointly formed a primary key, and a collection of *uncertain* attributes, over which a probability distribution was defined. Dey and Sarkar (1996) built a 1-NF representation of the probabilistic relations of Barbara et al. (1992) and described an extensive query algebra, which included such probabilistic database-specific operations as data compaction/ coalescence and conditionalization operations. The former, given two or more probability distributions for the same event, produces a *consensus* probability distribution. The latter computes the conditional probability distribution conditioned on a specific value of one or more of the uncertain attributes.

Zimanyi explicitly introduced possible worlds semantics of Nilsson to probabilistic databases (Zimanyi 1997). His framework uses the language of first-order probabilistic logic introduced by Halpern (1990). Zimanyi treats each probabilistic relation as a formula in the first-order probabilistic logic. He then defines full query algebra on by specifying how the formulas describing the probabilistic relations change when the operation is performed. While this approach is not very practical, it provides a clear semantics for query algebra operations.

In the mid- to late 1990s, a number of research groups extended probabilistic relational database frameworks. The work on the Semistructured Probabilistic Databases model (SPO) described in this chapter takes its roots from two such directions. The first is the work of Kornatzky and Shimony who proposed the first object-oriented probabilistic database framework (Kornatzky and Shimony 1994). This was, to our knowledge, the first extension of the work on probabilistic databases that extended beyond relational database model. In the framework of Kornatzky and Shimony, uncertainty was associated not just with the specific values of attributes, but also with the hierarchical structure of the objects themselves.

The second precursor to the work on the SPO model was ProbView, a relational probabilistic database management system which was the first framework to introduce *interval probabilities* to represent uncertainty in data. (Lakshmanan et al. 1997). ProvView stored the data in a compact, non-1-NF form, similar to the approach of Barbara et al. (1992). The semantics of the data though was defined using the 1-NF *annotated* probabilistic relations similar to those considered by Dey and Sarkar (1996). The query algebra operated on the annotated relations, so to answer a query ProbView translated the data into annotated form and performed the requisite operations.

In late 1990s, research on specific types of uncertainty in data appeared. Dyreson and Snodgrass (1998) considered temporal indeterminancy and introduced a probabilistic temporal database framework. Dekhtyar, Ross and Subrahmanian (2001) adopted and improved the ProbView approach to management of probabilities in temporal databases.

The SPO model (Dekhtyar et al. 2001, Zhao et al. 2005) came out of the observations that most of the probabilistic database frameworks at the time were not designed to store arbitrary probability distributions of multiple discrete random variables (with finite domains). A semistructured probabilistic object can store any probability distribution regardless of how many and which random variables are in it, what meta-data about the variables is present and known, and whether the probability distribution is conditional. The SPO framework was further relaxed by introducing interval probabilities to represent uncertainty (Goldsmith et al. 2003, Zhao et al. 2003, Zhao et al. 2004).

In parallel with our work on the SPO model, a number of alternative frameworks for management of uncertainty using semistructured data models and XML emerged. Hung et al. (Hung et al. 2003, Hung et al. 2003A) proposed Probabilistic XML framework (PIXML) to encode information about probability distribution. Nierman and Jagadish (2002) introduced ProTDB framework for the same purpose.

Some work addresses imprecise and uncertain data in sensor networks (Cheng et al. 2003, Deshpande et al. 2004, Tran et al. 2010, among others). The work presented in Sections 3 and 4 differs from the sensor network-based work in that we can process arbitrary query types based on the possible world semantics, but are not restricted to specific query operators. Note, however, that some of these approaches were shown to be more efficient (Tran et al. 2010). Independently, the MCDB project at University of Florida and IBM (Jampani et al. 2008) also employed Monte Carlo query processing for uncertain data, and focused on efficient integration of their techniques into a system. Previous work that is based on tuple uncertainty includes that by Dalvi and Suciu (2004) and by Benjelloun et al. (2006). In this chapter, our proposed techniques focus on attribute uncertainty, which is common in a number of application domains that we present.

Cheng et al. (2006) proposed probabilistic threshold join, which is similar to our v-join semantics. However, their query processing is based on x-bounds, which are a number of bounds for some data structure (e.g., each data page), unlike our dynamic filtering bounds using probability theory. In addition, we study indexing for efficient joins and a second type of semantics, d-join, which is useful for different applications.

Similarity join on data points in multidimensional space is well studied (e.g., Koudas and Sevcik 2000). The connection between this line of work and our work on d-join is due to the fact that we can reduce a d-join to a similarity join. However, when the dimensionality is high, with any existing technique, there is invariably a significant performance penalty. Our design of the condensed d-join and the optimal condensation scheme are a novel contribution. Dimensionality reduction is also studied for indexing time series databases (e.g., Keogh et al. 2001). However, a salient difference in discrete probability distributions than time series features is the constraint that probability values are between 0 and 1 and sum to 1. We take advantage of this and devise a simple, efficient, and optimal condensation algorithm. Finally, band-join on deterministic data was studied by DeWitt et al. (1991). V-join can deal with a variant of band-join on uncertain data, where the old techniques cannot be applied.

6 Conclusions

We have introduced several approaches to computation with probabilities, and given introductions to databases to support these approaches. The first approach, the Semistructured Probabilistic Objects framework, treats joint and conditional probability distributions as fundamental data objects. This supports reasoning with Bayesian networks, hidden Markov models (HMMs), and other probabilistic graphical models. We have discussed the queries possible with standard probability distributions, and mentioned some of the issues that arise when probability intervals are used. We also mention two implementations of the SPDBMS. The second approach, the tuple and attribute uncertainty framework, takes a data-centric approach and more tightly couples probability distributions with "data" itself. That is, either entities (tuples) or their properties (attributes) are extended with probabilities. We have discussed the semantics of general SQL queries, including joins, in this framework, and proposed some efficient query processing techniques.

We have presented a number of application scenarios in which significant uncertainty is present, and in which each of our approaches would be useful. We believe that a broad range of applications, albeit all containing uncertain information, would require database techniques tailored to their specific requirements.

As future work, it would be interesting to seamlessly integrate our approaches and produce an even more powerful system that can meet the diverse needs of modern applications.

References

Barbará, D., Garcia-Molina, H., Porter, D.: The Management of Probabilistic Data. IEEE Trans. Knowl. Data Eng. 4(5), 487–502 (1992)

Benjelloun, O., Das Sarma, A., Halevy, A., Widom, J.: ULDBs: Databases with Uncertainty and Lineage. In: VLDB (2006)

Bishop, C.: Pattern Recognition and Machine Learning. Springer (2007)

Block, C., Collins, J., Ketter, W.: Agent-based competitive simulation: Exploring future retail energy markets. In: Twelfth International Con-ference on Electronic Commerce, ICEC 2010, pp. 67–76. ACM (August 2010)

Brockwell, P., Davis, R.: Introduction to Time Series and Forecasting, 2nd edn. Springer Texts in Statistics (2002)

Burton, P., et al.: Size matters: just how big is BIG? – Quanti-fying realistic sample size requirements for human genome epidemiology. International Journal of Epidemiology 38, 263–273 (2009)

Cavallo, R., Pittarelli, M.: The Theory of Probabilistic Databases. In: VLDB, pp. 71–9 (1987)

de Campos, L.M., Huete, J.F., Moral, S.: Uncertainty Management Using Probability Intervals. In: Proc. International Conference on Information Processing and Management of Uncertainty (IPMU 1994), pp. 190–199 (1994)

Cheng, R., Kalashnikov, D., Prabhakar, S.: Evaluating probabilistic queries over imprecise data. In: SIGMOD (2003)

Cheng, R., Singh, S., Prabhakar, S., Shah, R., Vitter, J., Xia, Y.: Efficient Join Processing over Uncertain Data. In: CIKM (2006)

Dalvi, N., Suciu, D.: Efficient query evaluation on probabilistic databases. In: VLDB (2004)

Dekhtyar, A., Goldsmith, J., Hawkes, S.R.: Semistructured Probalistic Databases. In: Proc. SSDBM, pp. 36–45 (2001)

Dekhtyar, A., Ross, R.B., Subrahmanian, V.S.: Probabilistic temporal databases, I: algebra. ACM Trans. Database Syst. 26(1), 41–95 (2001)

Dekhtyar, A., Kevin Mathias, K., Gutti, P.: Structured Que-ries for Semistructured Probabilistic Data. In: Proc. 2nd Twente Data Manage-ment Workshop (TDM), pp. 11–18 (June 2006)

Deshpande, A., Guestrin, C., Madden, S., Hellerstein, J.M., Hong, W.: Model-driven data acquisition in sensor networks. In: VLDB (2004)

DeWitt, D., Naughton, J., Schneider, D.: An Evaluation of Non-Equijoin Algorithms. In: VLDB (1991)

Dey, D., Sarkar, S.: A Probabilistic Relational Model and Algebra. ACM Trans. Database Syst. 21(3), 339–369 (1996)

Dong, X., Halevy, A., Yu, C.: Data integration with uncer-tainty. The VLDB Journal (April 2009)

Dyreson, C.E., Snodgrass, R.T.: Supporting Valid-Time Indeterminacy. ACM Trans. Database Syst. 23(1), 1–57 (1998)

Ge, T.: Join Queries on Uncertain Data: Semantics and Efficient Processing. In: The Proceedings of the IEEE 27th International Conference on Data Engineering (ICDE 2011), Hannover, Germany (April 2011)

Ge, T., Li, Z.: Approximate Substring Matching over Uncertain Strings. The Proceedings of the VLDB Endowment (PVLDB Journal) 4(11), 772–782 (2011)

Ge, T., Zdonik, S.: Handling Uncertain Data in Array Database Systems. In: Proceedings of the IEEE 24th International Conference on Data Engineering (ICDE 2008), Cancun, Mexico (April 2008)

Goldsmith, J., Dekhtyar, A., Zhao, W.: Can Probabilistic Databases Help Elect Qualified Officials? In: Proceedings FLAIRS 2003 Conference, pp. 501–505 (2003)

Grimmett, G., Stirzaker, D.: Probability and Random Processes, 3rd edn. Oxford (2001)

Halpern, J.: An Analysis of First-order Logic of Probability. Artificial Intelligence 46(3), 311–350 (1990)

Hung, E., Getoor, L., Subrahmanian, V.S.: PXML: A Probabilistic Semistructured Data Model and Algebra. In: ICDE (2003)

Hung, E., Getoor, L., Subrahmanian, V.S.: Probabilistic Interval XML. In: ICDT 2003, pp. 358–374 (2003)

Jaffray, J.: Bayesian Updating and Belief Functions. IEEE Trans. on Systems, Man and Cybernetics 22(5), 1144–1152 (1992)

Jampani, R., Xu, F., Wu, M., Perez, L., Jermaine, C., Haas, P.: MCDB: A Monte Carlo Approach to Managing Uncertain Data. In: SIGMOD (2008)

Jestes, J., Li, F., Yan, Z., Yi, K.: Probabilistic String Similarity Joins. In: SIGMOD, pp. 327–338 (2010)

Keogh, E., Chakrabarti, K., Mehrotra, S., Pazzani, M.: Locally Adaptive Dimensionality Reduction for Indexing Large Time Series Databases. In: SIGMOD (2001)

Komatsu, K., et al.: Gene expression profiling following constitutive activation of MEK1 and transformation of rat intestinal epithelial cells. Molecular Cancer 5, 63 (2006)

Kornatzky, Y., Shimony, S.E.: A Probabilistic Object-Oriented Data Model. Data Knowl. Eng. 12(2), 143–166 (1994)

Koudas, N., Sevcik, K.: High Dimensional Similarity Joins: Algorithms and Performance Evaluation. In: TKDE (2000)

Lakshmanan, L.V.S., Leone, N., Ross, R.B., Subrahmanian, V.S.: ProbView: A Flexible Probabilistic Database System. ACM Trans. Database Syst. 22(3), 419–469 (1997)

Mann, M., Hendrickson, R., Pandey, A.: Analysis of Proteins and Proteomes by Mass Spectrometry. Annu. Rev. Biochem. 70, 437–473 (2001)

McDonald, M.: To Build a Better Grid. NY Times. July 28 (2011)

Mitzenmacher, M., Upfal, E.: Probability & Computing: Randomized Algorithms and Probabilistic Analysis. Cambridge U. Press (2005)

Nierman, A., Jagadish, H. V.: ProTDB: Probabilistic Data in XML. In: VLDB 2002, pp. 646–657 (2002)

Nilsson, N.J.: Probabilistic Logic. Artificial Intelligence 28(1), 71–87 (1986)

Ng, R., Subrahmanian, V.S.: Probabilistic Logic Programming. Inf. Comput. 101(2), 150–201 (1992)

Pearl, J.: Probabilistic Reasoning in Intelligent Systems: Networks of Plausible Inference. Morgan Kaufmann Publishers (1988)

Rosson, E.: Native XML Support for Semistructured Probabilistic Data Management, M.S. Thesis, Department of Computer Science, California Polytechnic State University (May 2008)

Szewczyk, R., et al.: An analysis of a large scale habitat monitoring application. In: SenSys (2004)

Tatbul, N., Buller, M., Hoyt, R., Mullen, S., Zdonik, S.: Confidence-based Data Management for Personal Area Sensor Networks. In: DMSN (2004)

Thiagarajan, A., Ravindranath, L., LaCurts, K., Mad-den, S., Balakrishnan, H., Toledo, S., Eriksson, J.: VTrack: Accurate, Energy-Aware Road Traffic Delay Estimation Using Mobile Phones. In: SenSys (2009)

Tran, T., Peng, L., Li, B., Diao, Y., Liu, A.: PODS: A New Model and Processing Algorithms for Uncertain Data Streams. In: SIGMOD (2010)

Walley, P.: Statistical Reasoning with Imprecise Probabilities. Chapman and Hall (1991)

Weichselberger, K.: The theory of interval-probability as a unifying concept for uncertainty. Int. J. Approx. Reasoning 24(2-3), 149–170 (2000)

Zhao, W., Dekhtyar, A., Goldsmith, J.: Query algebra operations for interval probabilities. In: Mařík, V., Štěpánková, O., Retschitzegger, W. (eds.) DEXA 2003. LNCS, vol. 2736, pp. 527–536. Springer, Heidelberg (2003)

Zhao, W., Dekhtyar, A., Goldsmith, J.: Databases for interval probabilities. Int. J. Intell. Syst. 19(9), 789–815 (2004)

Zhao, W., Dekhtyar, A., Goldsmith, J.: A Framework for Management of Semistructured Probabilistic Data. J. Intell. Inf. Syst. 25(3), 293–332 (2005)

Zimányi, E.: Query Evaluation in Probabilistic Relational Databases. Theor. Comput. Sci. 171(1-2), 179–219 (1997)

A Theoretically-Sound Approach for OLAPing Uncertain and Imprecise Multidimensional Data Streams

Alfredo Cuzzocrea

Abstract. In this chapter, we introduce a novel approach for tackling the problem of OLAPing uncertain and imprecise multidimensional data streams via novel theoretical tools that exploit probability, possible-worlds and probabilistic-estimators theories. The result constitutes a fundamental study for this exciting scientific field that, behind to elegant theories, is relevant for a plethora of modern data stream applications and systems that are more and more characterized by the presence of uncertainty and imprecision.

1 Introduction

Uncertain and *imprecise* data streams arise in a plethora of actual application scenarios ranging from *environmental sensor networks* to *logistic networks* and *telecommunication systems*, and so forth. Recently, research has devote a great deal of attention to this problem [9,21,2,10,28,23]. Consider, for instance, the simplest case of a sensor network monitoring the temperature $T$of a given geographic area W. Here, being T monitoring a natural, real-life measure, it is likely to retrieve an *estimate* of T, denoted by $\tilde{T}$, with a given *confidence interval*, denoted by $[\tilde{T}_{min}\tilde{T}_{max}]$, such that $\tilde{T}_{min} < \tilde{T}_{max}$, having a certain probability p_T, such that $0 \leq p_T \leq 1$, rather than to obtain the *exact value* of T, denoted by $\hat{T}$. The semantics of this confidence-interval-based model states that the (estimated) value of $T, \tilde{T}$, ranges between $\tilde{T}_{min}$ and $\tilde{T}_{max}$ with probability p_T. Also, a law describing the *probability distribution* according to which *possible values* of T vary over the interval $[\tilde{T}_{min}\tilde{T}_{max}]$ is assumed. Without loss of generality, the *normal distribution* is very often taken as reference. The normal distribution states that (possible) values in $[\tilde{T}_{min}\tilde{T}_{max}]$ have *all* the same probability to be the exact value of $T, \tilde{T}$, effectively. Despite the popularity of the normal distribution, the confidence-interval-based model above is prone to incorporate any other kind of probability distribution [25].

Alfredo Cuzzocrea
ICAR-CNR and University of Calabria, Italy
e-mail: cuzzocrea@sideis.unical.it

Z. Ma & L. Yan (Eds.): *Advances in Probabilistic Databases*, STUDFUZZ 304, pp. 109 – 129.
DOI: 10.1007/978-3-642-37509-5_5

While some recent papers have tackled the problem of efficiently representing, querying and mining uncertain and imprecise data streams [9,21,2,10,28,23], to the best of our knowledge, there not exist papers dealing with the problem of *efficiently OLAPing* [16] *uncertain and imprecise multidimensional data streams*, with explicit emphasis over multidimensionality of data. In order to fulfill this relevant gap, in this chapter we first introduce the problem of *estimating OLAP queries over uncertain and imprecise multidimensional data streams*, which can be reasonably considered as the first research attempt towards the definition of OLAP tools over uncertain and imprecise multidimensional data streams exposing *complete* OLAP functionalities, such as on-the-fly data summarization, indexing, and OLAM [17] primitives.

OLAP tools over multidimensional data streams have been recently regarded as one of the most attractive application scenarios for next-generation complex data stream management systems [3], due to specific characteristics of streaming data that are *inherently multidimensional, multi-level and multi-granularity in nature* [11,6,8,18]. From this breaking evidence, it has been widely recognized a *natural inadequateness* of conventional data stream mining tools [15] that do not adhere to a multidimensional and multi-resolution vision of data, which makes such tools not able of capturing the singular characteristics of streaming data above. In order to overcome this so-relevant limitation, several studies have proposed to apply "re-visited" versions of the traditional OLAP model to the peculiarities of data streams [3], by also integrating novel computational paradigms such as *intelligent techniques* [18], or *data compression* [11]. Unfortunately, these proposals do not consider the relevant problem of dealing with uncertainty and imprecision of data streams, which is instead investigated in this chapter.

According to motivations above, in this chapter we introduce the problem of answering OLAP queries over uncertain and imprecise multidimensional data streams, and propose a framework able of efficiently and effectively providing *theoretically-founded estimates* to such class of queries.

Figure 1 shows the reference application scenario of our research. In this scenario, a set of *multiple* data stream sources produce a set of multiple uncertain and imprecise multidimensional data streams. The final goal is that of supporting *on-the-fly* OLAP query evaluation over (uncertain and imprecise) multidimensional data stream *readings* in order to provide a *continuous answer* [4] to a *fixed* OLAP query for consumer data-stream-intensive applications and systems. Usually, data stream management systems make use of *buffering techniques* for query efficiency purposes [1], as consumer applications are very often interested in retrieving an *OLAP aggregation* (i.e., the result of an OLAP query) computed over a limited set of *recent* data stream readings [11] rather than querying the "history" of knowledge kept in data stream readings. This query model is also referred in literature as *window-based data stream query model* [1]. The solution consists in making use of a *buffer* storing an appropriately-selected collection of data stream readings (e.g., the last U, with $U > 0$). The problem of meaningfully determining the size of the buffer, denoted by B, is relevant in data stream research [1].

Typical OLAP aggregations of interest for consumer applications are: (*i*) SUM, which retrieves the summation of data stream readings stored in the buffer; (*ii*) COUNT, which retrieves the count of data stream readings stored in the buffer. On top of these baseline aggregations, *complex OLAP aggregations* can be devised [14].

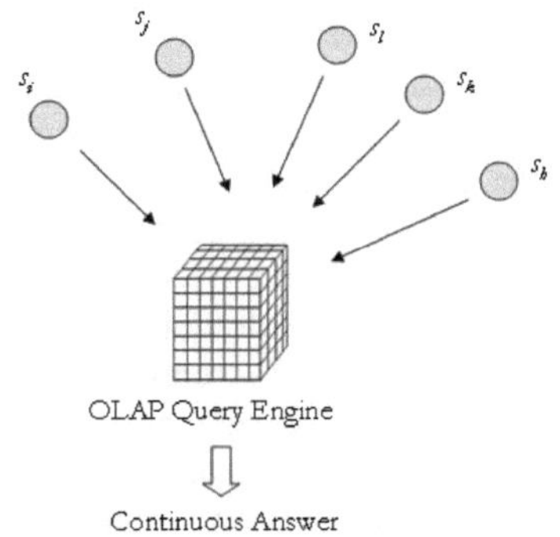

Fig. 1 OLAPing uncertain and imprecise multidimensional data streams

The solution to the problem of OLAPing uncertain and imprecise multidimensional data streams proposed in our research builds on some previous results that have been provided by recent research efforts. Particularly, [5], which focuses on static data, introduces a nice *Probability Distribution Function* (PDF) [25]-based model that allows us to capture the uncertainty of OLAP measures, whereas the imprecision of OLAP data with respect to *OLAP hierarchies* available in the multidimensional data stream model is meaningfully captured by means of the so-called *possible-world semantics* [5]. This semantics allows us to evaluate OLAP queries over uncertain and imprecise static data, while also ensuring some well-founded theoretical properties, namely *consistency*, *faithfulness* and *correlation-preservation* [5]. Similarly to the PDF-based model, the possible-world semantics is also exploited in our research, and specialized to the more challenging issue of dealing with uncertain and imprecise multidimensional data streams.

Summarizing, in this chapter we make the following three important contributions: (1) *we introduce a framework for supporting effective OLAP aggregations over uncertain and imprecise multidimensional data streams*; (2) *we provide an innovative possible-world semantics for OLAP aggregations over uncertain and imprecise multidimensional data streams*; (3) *we propose an innovative approach for providing theoretically-founded estimates to OLAP queries over uncertain and imprecise multidimensional data streams, via the probabilistic estimators theory.*

2 Modeling Uncertain and Imprecise Multidimensional Data Streams.

In this Section, we formally provide our proposed data model for uncertain and imprecise multidimensional data streams. This model relies-on and extends the research proposed in [5], which focuses on OLAP over static data. As highlighted in Section 1, [5] introduces a nice data model and a theoretically-sound possible-world semantics for OLAP over uncertain and imprecise static data, like those one can find in *probabilistic relational database systems* [13]. Particularly, [5] states that two specific occurrences of *data ambiguity* can arise in a typical OLAP scenario over static data: (*i*) uncertainty of (OLAP) measures, and (*ii*) imprecision of dimensional members of (OLAP) hierarchies.

We first introduce both these cases of data ambiguity in the particular context of dealing with multidimensional data streams by means of a meaningful example, and then formal definitions are provided.

Consider a large retail company selling computer components in USA throughout several branches located in different USA states. First, note that, due to the networked structure of the company, sale data sent by different branches to the company head-office periodically can be reasonable assumed as streaming data. Also, these data are clearly multi-dimensional in nature, according to several attributes that *functionally* model typical *sale facts*. Some example attributes are the following: (*i*) *Store*, which models the store where the sale has been performed; (*ii*) *Time*, which captures the time when the sale has been performed, (*iii*) *Product*, which models the sold product, (*iv*) *Customer*, which captures the customer that purchased the product. These attributes play the role of *OLAP dimensions* for the multidimensional data stream model. In addition to this, OLAP hierarchies are associated to these dimensions. An OLAP hierarchy describes the hierarchical relationship among *dimensional members* of the model [16]. Dimensional members are defined on top of functional attributes of the target OLAP analysis. For instance, consider the dimension *Store* of the running example. Here, a possible hierarchy associated to *Store* could be: *State* → *City* → *Store*. Possible instances of this hierarchy could be: (*i*) *California* → *Los Angeles* → *Computer Parts*; (*ii*) *California* → *San Francisco* → *PC Components*.

The main company could be interested in performing on-the-fly OLAP analysis of sales across the different branches. As a consequence, attribute *Sale*, which records the sale of a product p to a customer c performed in a certain store s during a certain day d, could be the *OLAP measure* of interest for the target analysis. Under a broader vision, performing OLAP analysis can be reasonably intended as executing a *set* of meaningful OLAP queries over multidimensional data streams. Therefore, goals of our research perfectly marry with the general requirement of achieving the definition of OLAP tools over uncertain and imprecise multidimensional data streams.

Due to data ambiguity, uncertainty of the measure *Sale* derives from the fact that data stream readings do not record the exact value of the sale ($e.g, 50\$$), but rather they record a confidence interval within which the possible value ranges ($e.g, [45,55]\$$) with a certain probability p_s, such that $0 \leq p_s \leq 1$.

Due to data ambiguity, imprecision of OLAP data is derives from the fact that data stream readings can record *any* of the dimensional members of the available hierarchies, at *any* hierarchical level, instead than leaf-level dimensional members *always* (as happens in the presence of *precise* OLAP data). Intuitively enough, data stream readings recording values at the leaf level of the whole OLAP hierarchy are defined as *precise readings* (i.e., the related OLAP level is univocally determined), whereas data stream readings recording values at any non-leaf level of the whole OLAP hierarchy are defined as *imprecise readings* (i.e., the related OLAP level is not univocally determined). For instance, in our running example target data stream readings could record the dimensional member *California*, so that it is not possible to precisely individuate the store where the related sale has been performed effectively. To give an example, focus the attention on the hierarchy instances above. The store could both be *Computer Parts*, or *PC Components*, alternatively. This is a source of imprecision when dealing with data stream readings produced by the target sources.

Inspired by [5], in our research we formally define an uncertain and imprecise *N*-dimensional data stream s as the tuple: $s = \langle ID_s, [V_{s,min}, V_{s,max}]p_s, t_s, a_{s,l,k_0}, a_{s,l,k_1} \dots, a_{s,l,k_{n-1}} \rangle$ such that:

- ID_s is the (absolute) identifier of s;
- $V_{s,min}$ is the lower bound of the confidence interval associated to the possible value of s;
- $V_{s,max}$ is the upper bound of the confidence interval associated to the possible value of s, such that $V_{s,min} < V_{s,max}$;
- p_s is the probability that the value of s ranges over the interval $[V_{s,min}, V_{s,max}]$;
- t_s is the timestamp in which s is recorded;
- $a_{s,l,k_0}, a_{s,l,k_1} \dots, a_{s,l,k_{n-1}}$, with $k_j \in \{0,1, \dots N-1\}$ and $n \leq N$, is a set of dimensional members associated to s, each one belonging to a certain OLAP hierarchy of the underlying *N*-dimensional data stream model.

Given a data stream s, $R_s = r_{s,0}, r_{s,1}, r_{s,2}, \dots$ denotes a(n) (unbounded) *sequence* of data stream readings (of s) $r_{s,j}$, such that $j \to \infty$.

Given a data stream model schema s like in the definition above, each data stream $r_{s,j}$ can be reasonably intended as an *instance* of s, i.e. a possible realization of s.

Given a data stream s and a buffer b having size $B > 0$, $R_S^B = r_{s,j}, r_{s,j+1}, r_{s,j+2}, \dots r_{s,j+B-1}$ denotes a B-bounded sequence of data stream readings (of s) $r_{s,j}$ stored within b, such that $|R_S^B| < B$.

It should be noted that, as highlighted in Section 1, many of today data stream management systems make use of buffers for query optimization purposes (e.g., [1]).

The information content of probabilistic measures of R_S^B can be nicely described by a *discrete* PDF P_S^B, defined as follows:

$$P_S^B = \sum_{k=0}^{B-1} \langle [v_{s,k,min}, v_{s,k,max}], p_{s,k} \rangle \cdot \delta[k] \qquad (1)$$

wherein: (*i*) $[v_{s,k,min}, v_{s,k,max}]$ denotes the confidence interval associated to the reading $r_{s,k}$ (of s); (*ii*) $p_{s,k}$is the probability that the value of $r_{s,k}$ ranges over the interval $[v_{s,k,min}, v_{s,k,max}]$; (*iii*) $\delta(\bullet)$ denotes the *Dirac impulse* [25].

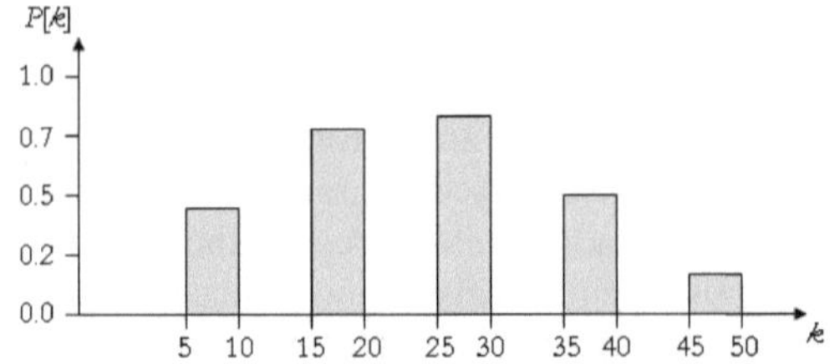

Fig. 2 A-5 bounded PDF for the OLAP measure *Sale* of the running example

To give an example, Figure 2 shows a 5-bounded PDF associated to the OLAP measure *Sale* of the running example. Here, $P_s^B[2]$ models the fact that the measure value ranges over the interval [25,30]$ with probability $p_2 = 0,5$, whereas $P_s^B[4]$ models the fact that the measure value ranges over the interval [45,50]$ with probability $p_4 = 0,1$.

3 Possible World Semantics for OLAPing Uncertain and Imprecise Multidimensional Data Streams

In this Section, we first review the possible-world semantics introduced in [5] for the case of OLAP over uncertain and imprecise static data, and then adapt this semantics to the more challenging case of streaming data. As mentioned in Section 1, we exploit such a possible-world semantics during OLAP query evaluation over uncertain and imprecise multidimensional data streams.

We review the possible-world semantics [5] via considering again our running example on the large retail company selling computer components in USA, introduced in Section 2. Figure 3 shows a two-dimensional OLAP view on multidimensional data streams generated by the different branches of the company. This view is exploited by the main company with the aim of making (on-the-fly) OLAP analysis of data streams via a set of OLAP queries posed against the view (see Section2). Particularly, the target OLAP view is built by simultaneously combining the dimensions *Store* and *Product* available in the whole multidimensional model of the case study, enriched by the respective hierarchies. A two-dimensional OLAP query posed against the view originates a continuous answer over multidimensional data streams (see Section 1).

For the sake of simplicity, for each hierarchy defined on the dimensions *Store* and *Product*, respectively, dimensional members of the leaf level are represented by white circles (see Figure 3), and the name of the corresponding dimensional attributes is not shown. Grey circles in Figure 3 represent instead data stream readings that populate the view. Recall that data stream readings arrive at different rates and arrival times, so that our running example is considering a *snapshot* of the OLAP view over multidimensional data streams at a certain time t_V.

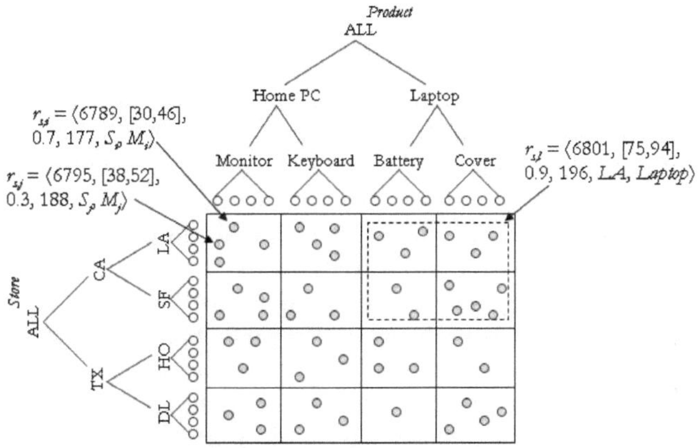

Fig. 3 The two dimensional OLAP view over uncertain and imprecise multidimensional data streams of the running example

Due to imprecision of OLAP hierarchies, data stream readings can record values at any level of OLAP hierarchies associated to the dimensions of the view (see Section 2). For instance, reading $r_{s,i} = \langle 6789, [30{,}46], 0.7, 177,\ S_i, M_i \rangle$ with identifier $ID_{s,i} = 6789$, which has been originated at timestamp $t_{s,i} = 177$, models a sale fact whose value ranges over the interval $[30{,}46]\$$ with probability $p_{s,i} = 0{,}7$. This fact is related to the sale of the monitor M_i performed in the store S_i. Therefore, this is a precise data stream reading (see Section 2). Similarly, data stream reading $r_{s,j} = \langle 6795, [38{,}52], 0.3, 188,\ S_j, M_j \rangle$ is precise as well.

Consider instead data stream reading $r_{s,l} = \langle 6801, [75{,}94], 0.9, 196, LA, Laptop \rangle$. At timestamp $t_{s,l} = 196$, this reading records a sale fact whose value ranges over the interval $[75{,}94]\$$ with probability $p_{s,l} = 0{,}9$. This fact is related to the sale of *one* product belonging to the (OLAP) group *Laptop* performed in *one* store located in *Los Angeles*. Therefore, this is an imprecise data stream reading (see Section 2) that introduces ambiguity in the target OLAP view.

"How to solve the ambiguity of reading $r_{s,l}$?" is a critical question in order to effectively estimate OLAP queries over uncertain and imprecise multidimensional data streams. To answer this question, first consider that reading $r_{s,l}$ can be related to *any* of the *possible* OLAP measures of the two-dimensional view contained within the two-dimensional range ⟨*California*, *Laptop*⟩ except the measures originated by precise data stream readings (i.e., the grey circles in Figure 3). In fact, the two-dimensional range ⟨*California*, *Laptop*⟩ stores measures related to sales of products belonging to the group *Laptop* performed in stores located in *California*.

[5] solves the problem above in the case of OLAP over uncertain and imprecise static data. First, explicitly note that in our running example the OLAP view at time t_V can be clearly considered as storing static data, i.e. data stream readings collected within the view until t_V. Therefore, in the following we refer to the latter static (sub-)case to review possible-world semantics introduced in [5]. According to [5], the ambiguity of $r_{s,l}$ can be solved by admitting the existence of a number of *different* possible worlds D_i built on top of the OLAP view, such that, in each

possible world D_i, $r_{s,l}$ is *alternatively* one of the measures of the two-dimensional view contained within ⟨*California*, *Laptop*⟩ that do not store a precise value. Since 13 measures stored in ⟨*California*, *Laptop*⟩ are precise, the number of possible worlds for the running example ($8 \times 8 - 13 = 64 - 13 = 51$). As stated in [5], the number of possible worlds over static OLAP data can become exponential. To become convinced of this, consider that: (*i*) our running example focuses on a simple two-dimensional OLAP view, whereas real-life corporate data cubes expose very high degrees of dimensionality [24]; (*ii*) the number of possible worlds has also a *combinatory dependence* on the depth of OLAP levels to which the imprecise value is referred to – our running example focuses on OLAP hierarchies having depth equal to 2, whereas real-life corporate data cubes are characterized by OLAP hierarchies having depths much higher than 2 [7]. In more detail, for k imprecise measures over a domain having C_j precise measures, we can have $\prod_{j=1}^{k} C_j$ possible worlds [5]. Also, a weight w_i is associated to each possible world D_i, in order to capture the likelihood of D_i of being the "true" world among the possible ones [5].

A critical aspect discussed in [5] is represented by the so-called *allocation policies*, which determine how to compute the probability to be assigned to an imprecise measure. This probability captures the likelihood of an imprecise measure of being one of the measures at the leaf-level aggregation of the target OLAP view. In this respect, several alternatives are proposed in [5]. As it will be clear throughout the chapter, similarly to the case of static data, computing probabilities of imprecise measures play a central role even in OLAP over uncertain and imprecise data, as, in turn, this heavily influences the query evaluation phase.

A first result of our research consists in stating that while the general possible-world semantics introduced in [5] is sound for uncertain and imprecise static data, *it cannot be applied to the context of uncertain and imprecise multidimensional data streams*. This mainly because, in the case of static data, the *entire* probabilistic database is *already available* so that the probabilistic OLAP framework can easily compute the *universe* of possible worlds. At a more practical level, this means that the overall space of possible worlds, although very large, can be somewhat reduced (e.g., in an asymptotic fashion). Obviously, this is not feasible in a data stream environment, where data stream readings arrive continuously, even at different rates, so that we cannot exploit the entire probabilistic reading repository in order to easily compute the universe of possible worlds. This makes the problem of dealing with the uncertainty and imprecision of multidimensional data streams much harder than the analogous static case.

4 A Data Driven Approach for Computing the Probabilities of Imprecise Multidimensional Data Streams Readings

In Section 3, we have clearly highlighted that computing the probabilities of imprecise multidimensional data stream readings is the basic issue to be faced-off in order to achieve the definition of a possible-world semantics for OLAP over uncertain and imprecise multidimensional data streams. Moreover, in Section 3 it has been put in emphasis that the approach proposed in [5] is not feasible for the case of streaming data.

Inspired by these main motivations, we propose an innovative approach to solve the relevant problem above, which is based on the well-known evidence stating that *OLAP data* (*static data as well as streaming data*) *are usually clustered and* (*highly*) *correlated in nature* [11,6,8,18] On the basis of this evidence, the main idea of the approach we propose consists in *computing the probability of the confidence interval of an imprecise multidimensional data stream reading to be "close" to confidence intervals of its neighboring precise multidimensional data stream readings*, fixed the multidimensional range of the imprecise multidimensional data stream reading, at a certain OLAP hierarchical level, as the reference neighborhood.

Consider a multidimensional OLAP view V over the target multidimensional data streams. Let S be the schema of the view V defined as the tuple: $S = \langle m, d_0, d_1, \dots d_{N-1} \rangle$, such that m is the measure of interest and $S = \langle d_0, d_1, \dots d_{N-1} \rangle$ the set of dimensions. Let $H = \langle h_0, h_1, \dots h_{N-1} \rangle$ be the set of hierarchies of S, such that h_i is the hierarchy associated to the dimension d_i. Also, let $Leaf(h_i)$ denotes the leaf level of dimensional members of the hierarchy h_i.

Consider the imprecise multidimensional data stream reading populating V, $r_{s,l} = \langle ID_{s,l}, [V_{s,l,min}, V_{s,l,max}]p_{s,l}, t_{s,l}, a_{s,l,k_0}, a_{s,l,k_1} \dots, a_{s,l,k_{n-1}} \rangle$, such that: (*i*) $k_j \in \{0,1, \dots N-1\}$, (*ii*) $n \leq N$, (*iii*) for each a_{s,l,k_j} in $r_{s,l}$, $a_{s,l,k_j} \notin Leaf(h_{k_j})$. We denote as $\langle a_{s,l,k_0}, a_{s,l,k_1}, \dots, a_{s,l,k_{n-1}} \rangle$ the multidimensional range associated to $r_{s,l}$, and as $\langle d_0, d_1, \dots d_{N-1} \rangle$ the *base* multidimensional range of the view V, which is obtained by combining *all* the dimensions of V at the leaf levels of their respective hierarchies. Since $a_{s,l,k_j} \notin Leaf(h_{k_j})$ for each a_{s,l,k_j} in $r_{s,l}$, $\langle a_{s,l,k_0}, a_{s,l,k_1}, \dots, a_{s,l,k_{n-1}} \rangle$ is *contained* by $\langle d_0, d_1, \dots d_{N-1} \rangle$, or, alternatively, $\langle a_{s,l,k_0}, a_{s,l,k_1}, \dots, a_{s,l,k_{n-1}} \rangle$ is a *proper sub-set* of $\langle d_0, d_1, \dots d_{N-1} \rangle$, i.e. $\langle a_{s,l,k_0}, a_{s,l,k_1}, \dots, a_{s,l,k_{n-1}} \rangle \subset \langle d_0, d_1, \dots d_{N-1} \rangle$. For the sake of simplicity, in the reminder we denote as $\langle a^{L_M}_{s,l,k_0}, a^{L_M}_{s,l,k_1}, \dots a^{L_M}_{s,l,k_{n-1}} \rangle$ the *projection* of $\langle a_{s,l,k_0}, a_{s,l,k_1}, \dots, a_{s,l,k_{n-1}} \rangle$ over $\langle d_0, d_1, \dots, d_{N-1} \rangle$, such that P_M is the depth of the whole OLAP hierarchy of V (i.e., the depth of the OLAP level of $\langle d_0, d_1, \dots d_{N-1} \rangle$). It should be noted that $\langle a^{L_M}_{s,l,k_0}, a^{L_M}_{s,l,k_1}, \dots a^{L_M}_{s,l,k_{n-1}} \rangle$ contains precise multidimensional data stream readings of the view V, being the latter defined at the leaf levels of hierarchies of the dimensions of V.

Let $p^{PW}_{r,l}$ denotes the *possible-world probability* to be assigned to the imprecise multidimensional data stream reading $r_{s,l}$. First, note that, since we deal with uncertain and imprecise data streams, $p^{PW}_{r,l}$ can be computed in terms of a *fraction* of the probability $p_{s,l}$ associated to $r_{s,l}$, i.e. $p^{PW}_{r,l} = \frac{p_{r,l}}{F}$, with $F > 0$. $p^{PW}_{r,l}$ captures the likelihood of $r_{s,l}$ of being a precise reading within $\langle a^{L_M}_{s,l,k_0}, a^{L_M}_{s,l,k_1}, \dots, a^{L_M}_{s,l,k_{n-1}} \rangle$.

Before introducing our approach for computing $p^{PW}_{r,l}$, the definition of *neighborhood* for an imprecise reading $r_{s,l}$, denoted by $N(r_{s,l})$, is necessary. Let $r_{s,l}$ be an imprecise reading, the multidimensional range associated to $r_{s,l}$, and Δ a positive integer parameter (i.e., $\Delta > 0$). Since $\langle a_{s,l,k_0}, a_{s,l,k_1}, \dots, a_{s,l,k_{n-1}} \rangle$ is an

n-dimensional domain, $r_{s,l}$ can be represented as an n-dimensional object in an n-dimensional space, as follows: $r_{s,l} = \langle r_{s,l,k_0}, r_{s,l,k_1}, \ldots, r_{s,l,k_{n-1}} \rangle$. Let $r_{s,l}^{L}$ denotes the projection of $r_{s,l}$ over$\langle a_{s,l,k_0}^{L_M}, a_{s,l,k_1}^{L_M}, \ldots a_{s,l,k_{n-1}}^{L_M} \rangle$. The Δ-*based neighborhood* of $r_{s,l}$ in$\langle a_{s,l,k_0}^{L_M}, a_{s,l,k_1}^{L_M}, \ldots a_{s,l,k_{n-1}}^{L_M} \rangle$, denoted by $\Delta\mathcal{N}_{\langle a_{s,l,k_0}^{L_M}, \ldots, a_{s,l,k_{n-1}}^{L_M} \rangle}(r_{s,l})$, is defined as the set of precise readings contained in $\langle a_{s,l,k_0}^{L_M}, a_{s,l,k_1}^{L_M}, \ldots a_{s,l,k_{n-1}}^{L_M} \rangle$ whose *Euclidean distance* from $r_{s,l}^{L}$ is lower or equal to Δ . Formally:

$$\Delta\mathcal{N}_{\langle a_{s,l,k_0}^{L_M}, \ldots, a_{s,l,k_{n-1}}^{L_M} \rangle}(r_{s,l}) = \left\{ \begin{array}{c} r'_{s,l} \; | r'_{s,l} \in \langle a_{s,l,k_0}^{L_M}, \ldots, a_{s,l,k_{n-1}}^{L_M} \rangle \\ r_{s,l} = \prod_{\langle a_{s,l,k_0}^{L_M}, \ldots, a_{s,l,k_{n-1}}^{L_M} \rangle}(r_{s,l}) \\ \sqrt{\sum_{i=0}^{n-1} \left(r'_{s,l,k_i} - r_{s,k_i}^{L} \right)^2} \le \Delta \end{array} \right\} \tag{2}$$

such that $\prod_D(p)$ denotes the projection operator of a multidimensional point p over a multidimensional domain D.

Following definition (2), we slightly modify the concept of possible-world probability $p_{r,l}^{PW}$, and introduce a variant that takes into account the Δ-based neighborhood of $r_{s,l}$, thus achieving the definition of the so-called Δ-*based possible-world probability*, denoted by $\Delta p_{r,l}^{PW}$. Obviously, when Δ= 1, then $p_{r,l}^{PW} \equiv \Delta p_{r,l}^{PW}$.

From Section 3, recall that a combinatory dependence between the depth of the OLAP hierarchical level of the multidimensional range $\langle a_{s,l,k_0}, a_{s,l,k_1}, \ldots, a_{s,l,k_{n-1}} \rangle$ associated to $r_{s,l}$ and computing $p_{r,l}^{PW}$, exists. We initially ignore this dependence and provide our proposed solution for computing $p_{r,l}^{PW}$, for the simplest case in which the multidimensional model of the view V is characterized by two (OLAP) levels only. This means that $\langle a_{s,l,k_0}, a_{s,l,k_1}, \ldots, a_{s,l,k_{n-1}} \rangle$ is *directly contained* by$\langle d_0, d_1, \ldots d_{N-1} \rangle$. Let P_L denotes the depth of the OLAP level of $\langle a_{s,l,k_0}, a_{s,l,k_1}, \ldots, a_{s,l,k_{n-1}} \rangle$ and P_M the depth of the OLAP level of $\langle d_0, d_1, \ldots d_{N-1} \rangle$, respectively. In other words, the property above is described by the following constraint:$P_M = P_L + 1$. After describing the proposed solution for the baseline case $P_M = P_L + 1$., we generalize this solution for OLAP views exposing an arbitrary number of (OLAP) levels.

Given an imprecise reading $r_{s,l}$ with multidimensional range $\langle a_{s,l,k_0}, a_{s,l,k_1}, \ldots, a_{s,l,k_{n-1}} \rangle$ and a precise reading $r'_{s,l}$ contained in$\langle a_{s,l,k_0}^{L_M}, a_{s,l,k_1}^{L_M}, \ldots, a_{s,l,k_{n-1}}^{L_M} \rangle$, we define the *probabilistic confidence interval distance* (*PCID*) of $r_{s,l}$with respect to $r'_{s,l}$, denoted by $PCID(r_{s,l}, r'_{s,l})$, as follows:

$$PCID(r_{s,l}, r'_{s,l}) = \frac{|V_{s,l,max} - V'_{s,l,max}|}{|V_{s,l,min} - V'_{s,l,min}|} \cdot p'_{s,l} \tag{3}$$

Intuitively enough, $PCID(r_{s,l}, r'_{s,l})$ is a *probabilistic factor* modeling how much the confidence interval of $r_{s,l}$ is "distant" from the confidence interval of $r'_{s,l}$. It should be noted that the more the quantity $\frac{|V_{s,l,max} - V'_{s,l,max}|}{|V_{s,l,min} - V'_{s,l,min}|}$ is low the more the absolute distance between the confidence interval of $r_{s,l}$ and the confidence interval of $r'_{s,l}$ is low. Also, since the confidence interval $[V_{s,l,min}, V_{s,l,max}]$ of $r'_{s,l}$ is a probabilistic estimate itself, such that $p'_{s,l}$ is the probability associated to it, the quantity $\frac{|V_{s,l,max} - V'_{s,l,max}|}{|V_{s,l,min} - V'_{s,l,min}|}$ must be multiplied by $p'_{s,l}$. This determines definition (3), and allows us to capture the "reliability" of $PCID(r_{s,l}, r'_{s,l})$ in modeling the probabilistic distance between the confidence interval of $r_{s,l}$, $[V_{s,l,min}, V_{s,l,max}]$ and the confidence interval of $r'_{s,l}$, $[V_{s,l,min}, V_{s,l,max}]$. Linearly, a low absolute distance involves in a low probabilistic distance, and vice-versa.

Given an imprecise reading $r_{s,l}$ with multidimensional range $\langle a^{L_M}_{s,l,k_0}, a^{L_M}_{s,l,k_1}, \ldots, a^{L_M}_{s,l,k_{n-1}} \rangle$ and its Δ-based neighborhood in $\langle a^{L_M}_{s,l,k_0}, a^{L_M}_{s,l,k_1}, \ldots, a^{L_M}_{s,l,k_{n-1}} \rangle$, $\Delta\mathcal{N}_{\langle a^{L_M}_{s,l,k_0}, \ldots, a^{L_M}_{s,l,k_{n-1}} \rangle}(r_{s,l})$, we define the *Δ-based neighborhood probabilistic confidence interval distance* ($\Delta\mathcal{N}PCID$) of $r_{s,l}$ with respect to $\Delta\mathcal{N}_{\langle a^{L_M}_{s,l,k_0}, \ldots, a^{L_M}_{s,l,k_{n-1}} \rangle}(r_{s,l})$, denoted by $\Delta\mathcal{N}PCID(r_{s,l}, \langle a^{L_M}_{s,l,k_0}, a^{L_M}_{s,l,k_1}, \ldots, a^{L_M}_{s,l,k_{n-1}} \rangle)$, as follows:

$$\Delta\mathcal{N}PCID\left(r_{s,l}, \langle a^{L_M}_{s,l,k_0}, a^{L_M}_{s,l,k_1}, \ldots, a^{L_M}_{s,l,k_{n-1}} \rangle\right) = \frac{\sum_{r'_{s,l} \in \Delta\mathcal{N}_{\langle a^{L_M}_{s,l,k_0}, \ldots, a^{L_M}_{s,l,k_{n-1}} \rangle}(r_{s,l})} PCID(r_{s,l}, r'_{s,l})}{COUNT_p\left(\Delta\mathcal{N}_{\langle a^{L_M}_{s,l,k_0}, \ldots, a^{L_M}_{s,l,k_{n-1}} \rangle}(r_{s,l})\right)} \quad (4)$$

such that $COUNT_p\left(\Delta\mathcal{N}_{\langle a^{L_M}_{s,l,k_0}, \ldots, a^{L_M}_{s,l,k_{n-1}} \rangle}(r_{s,l})\right)$ is an aggregate function that returns the number of precise readings in $\langle a^{L_M}_{s,l,k_0}, a^{L_M}_{s,l,k_1}, \ldots, a^{L_M}_{s,l,k_{n-1}} \rangle$.

Intuitively enough, $\Delta\mathcal{N}PCID\left(r_{s,l}, \langle a^{L_M}_{s,l,k_0}, a^{L_M}_{s,l,k_1}, \ldots, a^{L_M}_{s,l,k_{n-1}} \rangle\right)$ is a *probabilistic factor* modeling how much the confidence interval of $r_{s,l}$ is "distant" from confidence intervals of precise readings in the Δ-based neighborhood $\Delta\mathcal{N}_{\langle a^{L_M}_{s,l,k_0}, \ldots, a^{L_M}_{s,l,k_{n-1}} \rangle}(r_{s,l})$ (of $r_{s,l}$).

Upon the theoretical framework above, given an imprecise reading $r_{s,l}$ with multidimensional range $\langle a^{L_M}_{s,l,k_0}, a^{L_M}_{s,l,k_1}, \ldots, a^{L_M}_{s,l,k_{n-1}} \rangle$, the Δ-based possible-world probability of $r_{s,l}$, $\Delta p^{PW}_{r,l}$, is obtained as follows:

$$\Delta p^{PW}_{r,l} = \frac{p_{s,l}}{\Delta\mathcal{N}PCID\left(r_{s,l}, \langle a^{L_M}_{s,l,k_0}, a^{L_M}_{s,l,k_1}, \ldots, a^{L_M}_{s,l,k_{n-1}} \rangle\right)} \cdot \frac{1}{\sum_{r_{s,l} \in \langle a^{L_M}_{s,l,k_0}, \ldots, a^{L_M}_{s,l,k_{n-1}} \rangle} \Delta p^{PW}_{r,l}} \quad (5)$$

Intuitively enough, $\Delta p^{PW}_{r,l}$ is obtained in terms of a fraction of the probability $p_{s,l}$ associated to $r_{s,l}$, where the denominator is represented by the probabilistic factor ΔNPCID that *globally* takes into account the "distance" of the confidence interval of $r_{s,l}$ from confidence intervals of their neighboring precise readings in $\Delta\mathcal{N}_{\langle a^{L_M}_{s,l,k_0}, \ldots, a^{L_M}_{s,l,k_{n-1}} \rangle}(r_{s,l})$. Also, $\Delta p^{PW}_{r,l}$ is normalized with respect to all the other possible-world probabilities of imprecise readings in $\Delta\mathcal{N}_{\langle a^{L_M}_{s,l,k_0},\dots,a^{L_M}_{s,l,k_{n-1}}\rangle}(r_{s,l})$, in order to achieve a full probabilistic interpretation of the introduced theoretical setting.

When $\Delta = 1$, we revise definition (5) as follows:

$$p^{PW}_{r,l} = \frac{p_{s,l}}{\Delta\mathcal{N}PCID\left(r_{s,l},\langle a^{L_M}_{s,l,k_0},a^{L_M}_{s,l,k_1},\dots,a^{L_M}_{s,l,k_{n-1}}\rangle\right)|_{\Delta=1}} \cdot \frac{1}{\sum_{r_{s,l}\in\langle a^{L_M}_{s,l,k_0},\dots,a^{L_M}_{s,l,k_{n-1}}\rangle} p^{PW}_{r,l}} \tag{6}$$

Now focus the attention on the *extended* model for computing the possible-world probability of a given imprecise multidimensional data stream reading $r_{s,l}$ in the case that the multidimensional range $\langle a_{s,l,k_0}, a_{s,l,k_1}, \dots, a_{s,l,k_{n-1}}\rangle$ associated to $r_{s,l}$ is defined via combining n dimensions of the multidimensional model of V at *arbitrary* levels of their respective hierarchies, among those available in the OLAP hierarchy of V. For the sake of simplicity, assume that depths of these levels are equal for *all* the n hierarchies of V. Let P_L denote this common depth, and P_M the depth of $\langle d_0, d_1, \dots d_{N-1}\rangle$ in the OLAP hierarchy of V, respectively. Therefore, $P_M - P_L$ levels exist between $\langle a_{s,l,k_0}, a_{s,l,k_1}, \dots, a_{s,l,k_{n-1}}\rangle$ and $\langle d_0, d_1, \dots d_{N-1}\rangle$ in the OLAP hierarchy of V.

In order to compute the possible-world probability of $r_{s,l}$, $p^{PW}_{r,l}$, we simply *iterate* the baseline method given for a singleton pair of multidimensional ranges, i.e. $\langle a_{s,l,k_0}, a_{s,l,k_1}, \dots, a_{s,l,k_{n-1}}\rangle$ and $\langle d_0, d_1, \dots d_{N-1}\rangle$ of the previous case (where P_M and P_L satisfy the constraint $P_M = P_L + 1$), for *each* pair of multidimensional ranges defined at two *consecutive* levels of the OLAP hierarchy of V between P_L and P_M. In more detail, we pick pairs of multidimensional ranges across the OLAP hierarchy of V, starting from $\langle a_{s,l,k_0}, a_{s,l,k_1}, \dots, a_{s,l,k_{n-1}}\rangle$ at level P_L and until $\langle d_0, d_1, \dots d_{N-1}\rangle$ at level P_M is reached. Each pair of multidimensional ranges is constituted by the *actual* multidimensional range at level L_k, $\langle a^{L_k}_{s,l,k_0}, a^{L_k}_{s,l,k_1}, \dots, a^{L_k}_{s,l,k_{n-1}}\rangle$, and its *projection* over the multidimensional range at the underlying level L_{k+1}, $\langle a^{L_{k+1}}_{s,l,k_0}, a^{L_{k+1}}_{s,l,k_1}, \dots, a^{L_{k+1}}_{s,l,k_{n-1}}\rangle$. Therefore, for each pair of multidimensional ranges $\langle a^{L_k}_{s,l,k_0}, a^{L_k}_{s,l,k_1}, \dots, a^{L_k}_{s,l,k_{n-1}}\rangle$ and $\langle a^{L_{k+1}}_{s,l,k_0}, a^{L_{k+1}}_{s,l,k_1}, \dots, a^{L_{k+1}}_{s,l,k_{n-1}}\rangle$, a possible-world probability $p^{PW}_{r,l_{k,k+1}}$ is obtained. The final possible-world probability associated to$r_{s,l}$, $p^{PW}_{r,l}$, is obtained by multiplying *all* these probabilities, as follows:

$$p^{PW}_{r,l} = \prod_{k=P_L}^{P_M-1} p^{PW}_{r,l\;\,k,k+1} \tag{7}$$

Finally, it should be noted that our proposed approach for computing possible-world probabilities for imprecise multidimensional data stream readings above is completely suitable to deal with (tight) computational requirements posed by processing multidimensional data streams efficiently. In fact, computing these probabilities introduces low computational overheads, as, contrary to the baseline static case [5], a *reduced* number of precise multidimensional data stream readings only must be accessed to this end. According to the well-known general *locality principle*, these precise readings are supposed to be stored in the actual buffer used to answer OLAP queries over multidimensional data streams (see Section 1). Also, a significant difference between the possible-world semantics introduced by [5] for the case of OLAP over uncertain and imprecise static data and our revisited model for the more challenging case of OLAP over uncertain and imprecise multidimensional data streams can be recognized. In [5], being the entire probabilistic database already available, the whole universe of possible worlds can be computed, by also exploiting somewhat asymptotic approximation. The same cannot be for the research context we investigate, as the solution [5] for the static case would introduce a combinatorial explosion in the dynamic case as, for *each* new multidimensional data stream reading arrival, the universe of possible worlds should be (re-)computed. Then, after each arrival this huge collection of possible worlds should be summarized. Computational overheads introduced by the latter solution would make the method [5] not suitable to be integrated within the core layer of OLAP tools over uncertain and imprecise multidimensional data streams. Contrary to this, our proposed solution exploits a data-driven approach that, while being still sound and compliant with the fundamental possible-world semantics, it *locally* computes a *unique* possible world for each new multidimensional data stream reading arrival, thanks to nice abstractions are founded on the clustered and highly-correlated nature of OLAP data, and the innovative multidimensional neighborhood concept. This contributes to significantly tame the computational overheads introduced by our proposed solution, which is thus perfectly prone to be integrated within the core layer of OLAP tools over uncertain and imprecise multidimensional data streams.

5 Background: Probabilistic Estimators for Database Management

In this Section, we provide the background knowledge on *probabilistic estimators* [25], which we exploit in our framework in order to provide accurate estimates to OLAP queries over uncertain and imprecise multidimensional data streams.

Probabilistic estimators constitutes a well-known theory in the context of *theoretical statistics* [25]. Basically, given a parameter p to be measured, probabilistic estimators deal with the issue of providing an *accurate estimate* of p, $\tilde{p}$, supposed that the exact value of p, $\hat{p}$, cannot be retrieved with sufficient precision or, at the same, it is not possible at all to retrieve $\hat{p}$. To give an example, p could be an attribute of a relational table T_P stored in a probabilistic database D_P , and the measure could be the answer to a conventional SQL query Q involving p. Since attribute values in p are probabilistic in nature, the set of exact tuples representing the answer to Q cannot be retrieved so that the alternative strategy is represented by obtaining reliable estimates on the values of these tuples. This simple yet effective theory has been widely used in database research, and particularly in the context of a plethora of applications ranging from estimate-based query answering techniques over probabilistic databases [26] and data streams [9,21] to *approximate query answering techniques* over large databases [19] and data cubes [12,22], and so forth.

In order to build a reliable probabilistic estimator over p, a *random variable* X_p must be introduced. The domain of X_p, denoted by $D(X_p)$, is composed by a *finite* set of (probabilistic) *events*, whereas the co-domain of X_p, denoted by $CD(X_p)$, represents a possible value of the observed parameter p related to the (probabilistic) occurrence of a certain event e, i.e. $X_p(e) = \tilde{p}$, or, similarly, an *instance* of X_p, to which a certain probability p_e is associated. Upon the random variable X_p, a probabilistic estimator for the parameter p, denoted by $\Psi(X_p)$, is introduced, and exploited to retrieve an estimate over p as related to the occurrence of a certain event e, i.e. $X_p(e) = \tilde{p}$. The way of defining the random variable X_p determines the *class* of the probabilistic estimator $\Psi(X_p)$. According to this general theoretical framework, a wide family of probabilistic estimators, each one meant for a particular application domain, exists in literature [25].

For instance, the *Hoeffding*-based estimator [19], denoted by $\Psi_H(X_p)$, has been extensively used in past research efforts with the goal of providing approximate answers (based on accurate estimates) to queries over large databases [19] and data cubes [12,22]. $\Psi_H(X_p)$ is based on the well-known *Hoeffding's inequality* [20], which asserts the following. Given (*i*) a set of Z *independent* random variables $\chi_p = \{X_{p,0}, X_{p,1}, \dots X_{p,Z-1}\}$, (*ii*) a scalar $r \geq 0$ bounding random variables in χ_p, i.e. $0 \leq X_{p,z} \leq r$ for each $z \in \{0, 1, \dots, Z-1\}$, for each $\varepsilon > 0$ the following inequality holds:

$$P\left(|\overline{X}_p - \mu_p| \leq \varepsilon\right) \geq 1 - 2 \cdot e^{\frac{2 \cdot Z \cdot \varepsilon^2}{r^2}} \tag{8}$$

wherein χ_p is the *sample mean* of χ_p, and μ_p is the average value of the observed instance of χ_p, denoted by$\{\chi_p(e_0) = \tilde{p}_0, \chi_p(e_1) = \tilde{p}_1, \dots, \chi_p(e_{z-1}) = \tilde{p}_{z-1}\}$. The Hoeffding-based estimator belongs to the wide family of $\langle\varepsilon, \delta\rangle$-based probabilistic estimators, which assert to produce an estimate $\tilde{X}$ of a random variable X with probability $P(|\tilde{X} - X| \leq \varepsilon) \geq 1 - \delta$, such that $\varepsilon > 0$ and $\delta > 0$ are arbitrarily small.

Hoeffding's inequality (8) allows us to obtain an accurate estimate on the average value of the observed instance of χ_p, $\{X_p(e_0) = \tilde{p}_0, X_p(e_1) = \tilde{p}_1, \dots, X_p(e_{z-1}) = \tilde{p}_{z-1}\}$, whose *estimate error* is even probabilistically bounded by (8). This gives us a reliable solution to obtain a probabilistically bounded approximate answer to AVG-based queries over the instance of χ_p, $\{X_p(e_0) = \tilde{p}_0, X_p(e_1) = \tilde{p}_1, \dots, X_p(e_{z-1}) = \tilde{p}_{z-1}\}$.

6 Supporting OLAP Query Evaluation Over Uncertain and Imprecise Multidimensional Data Stream via Reliable Probabilistic Estimators

In this Section, we provide definition and fundamental features of our proposed innovative *$\langle\varepsilon, \delta\rangle$-based probabilistic estimator* [25] that retrieves estimates to

OLAP queries via *appropriate statistics* extracted from PDF describing the uncertain and imprecise multidimensional data stream readings.

Given a set of multiple uncertain and imprecise multidimensional data stream readings populating the reference M-dimensional OLAP view $V$and a fixed input OLAP query Q over V, our final goal is to provide an accurate estimate to Q, $\tilde{Q} \cdot Q$. can be modeled by the following tuple: $Q = \langle \mathcal{A}, R_{k_0}, R_{k_1}, \dots, R_{k_{m-1}} \rangle$, such that (*i*) $\mathcal{A}$ is an SQL aggregate operator (e.g., $SUM, COUNT$, etc), and (*ii*) R_{k_i} is a *dimensional range* over the M-dimensional model of V, where $k_i \in \{0,1,\dots,M-1\}$and $m \leq M$.

The OLAP query definition above can be slightly modified in order to achieve the definition of a *continuous OLAP query* over V, denoted as $Q(t) = \langle Q, t \rangle$, such that Q is a conventional OLAP query over V (defined like in the previous case) and t is a timestamp in which the fixed query Q is evaluated against V. By iterating the query task above for a set of timestamps $T_Q = \{t_q, t_{q+1}, \dots, t_{q+k}\}$, we finally obtain a set of queries $Q_T = \left\{Q_{(t_q)}, Q_{(t_{q+1})}, \dots, Q_{(t_{q+k})}\right\}$ and a set of estimates $\tilde{Q}_T = \left\{\tilde{Q}_{(t_q)}, \tilde{Q}_{(t_{q+1})}, \dots, \tilde{Q}_{(t_{q+k})}\right\}$, such that $\tilde{Q}_{(t_q)}$ is the estimate to the query $Q(t_q) = \langle Q, t_q \rangle$ at timestamp $t_q \cdot \tilde{Q}(t)$ represents the continuous answer to the continuous query $Q(t)$.

For the sake of simplicity, in the following we focus the attention on the baseline problem of providing the accurate estimate $\tilde{Q}$, due to the fact that providing the continuous answer $\tilde{Q(t)}$ to the continuous query $Q(t_q)$ can be straightforwardly obtained via simply iterating the baseline solution.

From Section 1, recall that, similarly to state-of-the-art data stream query processing techniques, in our framework we make use of a buffer b of size B to query efficiency purposes. Therefore, in our reference application scenario (see Section 1), consumer applications are interested in executing OLAP queries over the collection of data stream readings stored within b.

Due to uncertainty an imprecision of multidimensional data stream readings, a probabilistic estimator $\Psi(Q)$ must be introduced in order to provide an accurate estimate to Q, $\tilde{Q}$. To this end, our general approach consists in providing $\tilde{Q}$ in terms of *some statistics* extracted from an *appropriate* PDF describing uncertain and imprecise multidimensional data stream readings [9,21]. Thanks to the probabilistic data stream model introduced in Section 2 and the possible-world semantics for OLAP over uncertain and imprecise multidimensional data streams introduced in Section 4, *we achieve the definition of a reliable* $\langle \varepsilon, \delta \rangle$ *−based probabilistic estimator over uncertain and imprecise multidimensional data streams*, denoted by $\langle \varepsilon, \delta \rangle - \Psi(Q)$. Therefore, since the theory for probabilistic estimators is already available and it can be directly exploited within our proposed framework for OLAP over uncertain and imprecise multidimensional data streams, the most important research contribution we provide in this respect is represented by the *methodology for building the appropriate PDF describing uncertain and imprecise multidimensional data stream readings.* We next discuss this aspect, which plays a critical role in our research.

First, note that, in classical probabilistic estimators like the Hoeffding-based one, the PDF describing the set of target random variables X_p, denoted by P_p, *is not available.* In fact, the underlying goal of the Hoeffding's inequality consists in "re-constructing" this unknown PDF in order to retrieve probabilistic bounds over estimates of observed parameters. Now focus the attention on probabilistic properties of our proposed framework for OLAP over uncertain and imprecise multidimensional data streams. For each precise data stream reading $r'_{s,l}$ thanks to our probabilistic data stream model (see Section 2), the B-bounded discrete PDF describing the variation of possible values of $r'_{s,l}$ in terms of confidence intervals, $P^B_{r'_{s,l}}$, is available. For each imprecise data stream reading $r_{s,l}$, thanks to our possible-world semantics for OLAP over uncertain and imprecise data stream (see Section 4), a B-bounded discrete PDF accomplishing the same task illustrated for the case of *precise* data stream readings, $P^B_{r'_{s,l}}$, can be computed. This allows us to achieve a *unifying approach* for treating both precise and imprecise data stream readings, respectively. Therefore, for the sake of simplicity, in the remainder of this chapter we simple refer to multidimensional data stream readings $r_{s,l}$ (for both precise and imprecise instances) and related PDF, $P^B_{r_{s,l}}$.

The nice unifying PDF-based probabilistic data model above allows us to *directly retrieve* an accurate estimate $\tilde{Q}$ to an input OLAP query Q over the multidimensional view V via appropriate statistics extracted from the PDF describing the involved uncertain and imprecise multidimensional data stream readings. This is a significant contribution over the methodology proposed in [5], which is tailored to static OLAP data only.

For the sake of simplicity, consider SUM-based OLAP queries over uncertain and imprecise multidimensional data streams as the reference class of queries we deal with. Note that other kinds of OLAP queries (such asCOUNT-based, AVG-based etc) can be easily obtained on top of SUM-based ones.

Given a SUM-based OLAP query $Q = \langle SUM, R_{k_0}, R_{k_1}, \ldots, R_{k_{m-1}} \rangle$ over the M-dimensional OLAP view V populated by precise and imprecise multidimensional data stream readings, our approach for providing a theoretically-founded estimate to Q, $\tilde{Q}$, is based on the following 3-step methodology:

1. On the basis of the multidimensional selectivity of Q, obtained by combining the dimensional ranges $R_{k_0}, R_{k_1}, \ldots, R_{k_{m-1}}$ into the multidimensional range $\langle R_{k_0}, R_{k_1}, \ldots, R_{k_{m-1}} \rangle$, retrieve the set of J PDF associated to data stream readings of V involved by Q, denoted by $P^B_Q = \left\{P^B_{r_{s,l,0}}, P^B_{r_{s,l,1}}, \ldots, P^B_{r_{s,l,J-1}}\right\}$ – this can be simply performed via checking, for each reading $r_{s,l}$, if *all* its dimensional members $a_{s,l,k_0}, a_{s,l,k_1}, \ldots, a_{s,l,k_{m-1}}$ (see Section 2) are contained within $\langle R_{k_0}, R_{k_1}, \ldots, R_{k_{m-1}} \rangle$, i.e. $a_{s,l,k_j} \subset \langle R_{k_0}, R_{k_1}, \ldots, R_{k_{m-1}} \rangle$ for each $k_j$in$\{0,1,\ldots,M-1\}$.
2. From the set of PDF P^B_Q, compute the *joint PDF* [25], denoted by $\mathcal{G}^B_Q$ – intuitively enough, $\mathcal{G}^B_Q$ models the joint contribution of *all* the PDF associated to data stream readings involved by Q.

3. Retrieve an accurate estimate to Q, $\tilde{Q}$, as follows:

$$\tilde{Q} = \int_{\langle R_{k_0}, R_{k_1}, \ldots, R_{k_{m-1}} \rangle} \mathcal{G}_Q^B\,[k]dk \tag{9}$$

Since step 1 is trivial and it does not deserve particular attention, whereas steps 2 and 3 are more significant, we next explode and discuss the latter steps.

Step 2 computes the joint PDF from the PDF set P_Q^B, $\mathcal{G}_Q^B$. The latter is the main contribution of our research regarding to query uncertain and imprecise data streams. To this end, we propose an approach inspired by [27], where PDF describing independent observations are combined for *pre-aggregation* purposes in the context of *probabilistic Data Warehouse servers*.. According to our proposed approach, given two PDF $P_{r_{s,l_i}}^B$ and $P_{r_{s,l_j}}^B$ describing the data stream readings r_{s,l_i} and r_{s,l_j}, respectively, the joint PDF $\mathcal{G}_{i,j}^B$ can be obtained as follows. For each component k of $\mathcal{G}_{i,j}^B$, the pair of confidence intervals $[V_{s,l,min_i}, V_{s,l,max_i}]$. and $[V_{s,l,min_j}, V_{s,l,max_j}]$ of $P_{r_{s,l_i}}^B$[k] and $P_{r_{s,l_j}}^B$[k], respectively (picked at the *same* buffer entry k), are used to compute the *new* confidence interval $[V_{s,l,min_i} + V_{s,l,min_j}, V_{s,l,max_i} + V_{s,l,max_j}]$. Similarly, the pair of probabilities p_{s,l_i} and p_{s,l_j} of $P_{r_{s,l_i}}^B$[k] and $P_{r_{s,l_j}}^B$[k], respectively (picked at the *same* buffer entry k), are then multiplied in order to obtain the *new* probability $p_{s,l_i} \cdot p_{s,l_j}$ The k-th component of $\mathcal{G}_{i,j}^B$ is finally obtained as follows:

$$\mathcal{G}_{i,j}^B[k] = \langle [V_{s,l,min_i} + V_{s,l,min_j}, V_{s,l,max_i} + V_{s,l,max_j}], p_{s,l_i} \cdot p_{s,l_j} \rangle\,.$$

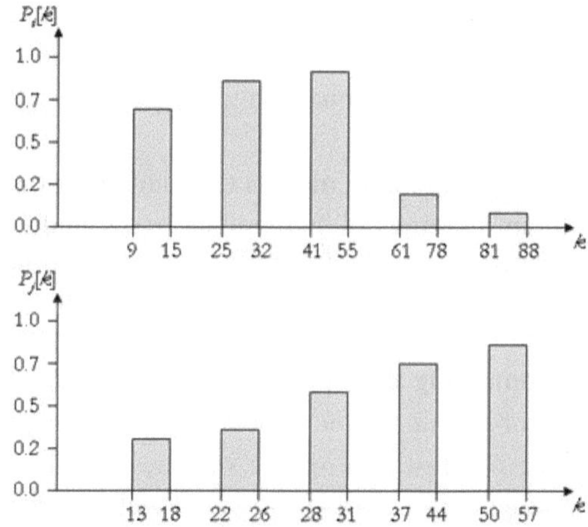

Fig. 4 Two 5-bounded PDF for the two data streams readings $\boldsymbol{r}_{s,l_i}$ and $\boldsymbol{r}_{s,l_j}$ of the running example

Figure 4 shows two 5-bounded PDF $P_{r_{s,l_i}}^B$ and $P_{r_{s,l_j}}^B$ associated to two data stream readings r_{s,l_i} and r_{s,l_j}, respectively. In particular, $P_{r_{s,l_i}}^B$ is defined as follows:

$$P^B_{r_{s,l_i}}[\mathrm{k}] = \langle[13,18],0.68\rangle \cdot \delta(0) + \langle[22,26],0.85\rangle \cdot \delta(1) + \langle[28,31],0.94\rangle \cdot \delta(2) + \langle[37,44],0.18\rangle \cdot \delta(3) + \langle[50,57],0.04\rangle \cdot \delta(4)$$

whereas $P^B_{r_{s,l_j}}$ is defined as follows:

$$P^B_{r_{s,l_i}}[\mathrm{k}] = \langle[9,15],0.26\rangle \cdot \delta(0) + \langle[25,32],0.33\rangle \cdot \delta(1) + \langle[41,55],0.59\rangle \cdot \delta(2) + \langle[61,78],0.68\rangle \cdot \delta(3) + \langle[81,88],0.75\rangle \cdot \delta(4)$$

On the basis of the above-described approach, the joint PDF computed from $P^B_{r_{s,l_i}}$ and $P^B_{r_{s,l_j}}$, $\mathcal{G}^B_{i,j}$, which is shown in Figure 5, is obtained as follows:

$$\mathcal{G}^B_{i,j}[k] = \langle[22,33],0.16\rangle \cdot \delta(0) + \langle[47,58],0.27\rangle \cdot \delta(1) + \langle[69,86],0.64\rangle \cdot \delta(2) + \langle[98,122],0.11\rangle \cdot \delta(3) + \langle[131,145],0.03\rangle \cdot \delta(4)$$

Step 3 retrieves an accurate estimate to Q, $\tilde{Q}$, by means of the *multidimensional integral* (9), which models a conceptual formula showing how $\tilde{Q}$ can be retrieved, at a theoretical level, in the *continuous domain*. Recall that we handle SUM-based OLAP queries over uncertain and imprecise multidimensional data stream readings, so that the integral operator (9) well-defines $\tilde{Q}$ as the summation of partial contributions over the continuous domain. Since we focus on discrete PDF (see Section 2), (9) is then specialized for the *discrete domain*, as follows:

$$\tilde{Q} = \sum_{\langle k_0,k_1,\ldots,k_{m-1}\rangle=\langle R_{k_0}[0],R_{k_1},R_{k_{m-1}}[0]\rangle}^{\langle R_{k_0}[|R_{k_0}|-1],R_{k_1}[|R_{k_1}|-1],\ldots,R_{k_{m-1}}[|R_{k_{m-1}}|-1]\rangle} \mathcal{G}^B_Q[\langle k_0,k_1,\ldots,k_{m-1}\rangle] \quad (10)$$

such that *(i)* $\langle k_0,k_1,\ldots,k_{m-1}\rangle$ denotes an m-dimensional point, *(ii)* $R_{k_i}[h]$ the h-th dimensional member of the dimensional range R_{k_i}, and *(iii)* $|R_{k_i}|$ the cardinality of R_{k_i}.

Furthermore, since *(i)* $\mathcal{G}^B_Q$ globally models the joint contribution of all the PDF associated to data stream readings of V involved by Q, and *(ii)* the multidimensional selection performed at step 1 of our OLAP query estimation technique selects *only* the set of J PDF associated to the involved readings, we can *break* the dependency of $\mathcal{G}^B_Q$ from the m-dimensional component exposed in (10), and retrieve $\tilde{Q}$ by means of computing appropriate statistics over $\mathcal{G}^B_Q$. From the literature, relevant *moments* [25] of $\mathcal{G}^B_Q$, i.e. *mean*, denoted by $\boldsymbol{E}(\mathcal{G}^B_Q)$, and *variance*, denoted by $\boldsymbol{V}(\mathcal{G}^B_Q)$, play the role of most appropriate statistics in this respect. In fact, according to several studies [9,21], these moments well-summarize the *statistical content* of $\mathcal{G}^B_Q$, and are well-suited for OLAP queries over the involved (uncertain and imprecise) readings. For instance, the estimate to a SUM-based OLAP query can be obtained as follows:$\tilde{Q} = B \cdot \boldsymbol{E}(\mathcal{G}^B_Q)$, where B denotes the buffer size (see Section 1), $\boldsymbol{E}(\mathcal{G}^B_Q)$the mean of $\mathcal{G}^B_Q$.

Given a discrete PDF $P[k]$with q samples, the mean $\boldsymbol{E}(P[k])$of $P[k]$ is defined as follows [25]:

$$(P[k]) = \sum_{i=0}^{q-1} k_i \cdot P[k_i] \tag{11}$$

whereas the variance $V(P[k])$of $P[k]$ is defined as follows [25]:

$$\boldsymbol{V}(P[k]) = \boldsymbol{E}([P(k_i - \mu)^2]) = \sum_{i=0}^{q-1}(k_i - \mu)^2 \cdot P[k_i] \tag{12}$$

To give an example, consider the joint PDF $\mathcal{G}_{i,j}^B$ of Figure 5. Here:

$$\boldsymbol{E}(\mathcal{G}_Q^B) = \begin{array}{l} [22 \cdot 0.16, 33 \cdot 0.16] + [47 \cdot 0.27, 58 \cdot 0.27] + [69 \cdot 0.64, 86 \cdot 0.64] \\ + [98 \cdot 0.11, 122 \cdot 0.11] + [131 \cdot 0.03, 145 \cdot 0.03] \\ = [3.52, 5.28] + [12.69, 15.66] + [44.16, 55.04] \\ + [10.78, 13.42] + [3.93, 4.35] = [75.08, 93.75] \end{array} \tag{13}$$

and:

$$\boldsymbol{V}(\mathcal{G}_Q^B) = \begin{array}{l} [(22 - 75.08)^2 \cdot 0.16, (33 - 93.75)^2 \cdot 0.16] \\ + [(47 - 75.08)^2 \cdot 0.27, (58 - 93.75)^2 \cdot 0.27] \\ + [(69 - 75.08)^2 \cdot 0.64, (86 - 93.75)^2 \cdot 0.64] \\ + [(98 - 75.08)^2 \cdot 0.11, (122 - 93.75)^2 \cdot 0.11] \\ + [(131 - 75.08)^2 \cdot 0.03, (145 - 93.75)^2 \cdot 0.03] \\ = [450.80, 590.49] + [212.88, 345.08] + [23.66, 38.44] \\ + [57.78, 87.79] + [98.80, 78.80] = [843.92, 1140.60] \end{array} \tag{14}$$

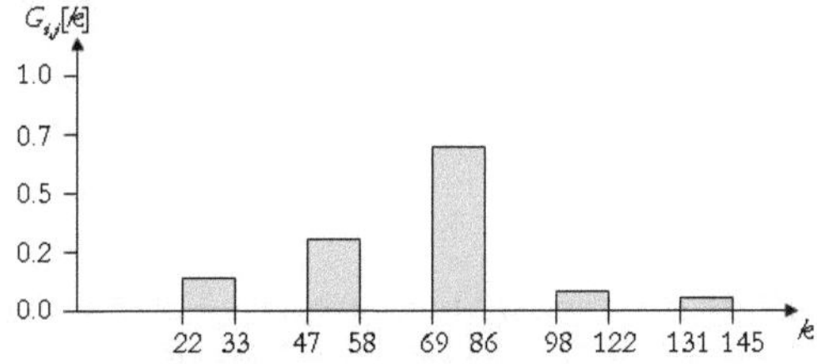

Fig. 5 The bounded joint PDF computed from the two PDF of Figure 4

Thanks to the statistical framework above, we are able of providing confidence intervals for both mean of $\mathcal{G}_Q^B$, $\boldsymbol{E}(\mathcal{G}_Q^B)$ (i.e.,[75.08,93.75]), and variance of $\mathcal{G}_Q^B$, $\boldsymbol{V}(\mathcal{G}_Q^B)$ (i.e.,[843.92, 1140.60]), respectively.

How to compute the parameters ε and δ of the probabilistic estimator $\langle\varepsilon, \delta\rangle - \Psi(Q)$ for statistics $\boldsymbol{E}(\mathcal{G}_Q^B)$and $\boldsymbol{V}(\mathcal{G}_Q^B)$? Since $\mathcal{G}_Q^B$ is a joint PDF obtained from a set of PDF, it can be supposed that $\mathcal{G}_Q^B$ follows a *quasi-Gaussian distribution* [25] (see Figure 6). Therefore, we can exploit *numerical methods* to compute both parameters ε and δ. For each probability $p = 1 - \delta$,which is equal to the area of the Gaussian bell, fixed the confidence interval $[\tilde{X} - \varepsilon, \tilde{X} + \varepsilon]$ of an estimate $\tilde{X}$, we can retrieve the corresponding value of the parameter ε, such that$P(|\tilde{X} - X| \leq \varepsilon) \geq 1 - \delta$. ε is recognized as the *accuracy* of the probabilistic estimator $\langle\varepsilon, \delta\rangle$-$\Psi(Q)$ for *that* estimate $\tilde{X}$ [25].

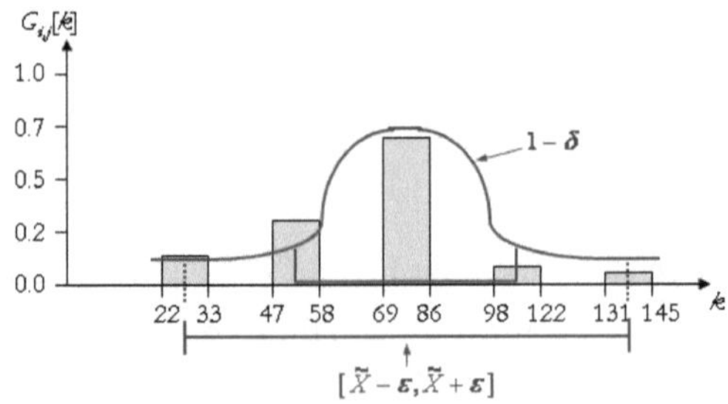

Fig. 6 The quasi-Gaussian Distribuition built on top of the 5-bounded joint PDF of figure 5

7 Conclusions and Future Work

In this chapter, we have proposed a novel framework for estimating OLAP queries over uncertain and imprecise multidimensional data streams. The proposed framework introduces some relevant research contributions, and the suitability of the framework in the context of modern data stream applications and systems, which are more and more characterized by the presence of uncertainty and imprecision, has been demonstrated throughout several reliable study cases.

Future work is mainly oriented towards making our proposed framework robust with respect to complex OLAP aggregations over uncertain and imprecise multidimensional data streams, according to research directions described in [14], and beyond simple SQL-based OLAP aggregations (e.g., SUM, COUNT etc) like those investigated in this chapter.

References

1. Abadi, D., Carney, D., Cherniack, M., Convey, C., Lee, S., Stonebraker, M., Tatbul, N., Zdonik, S.: Aurora: A New Model and Architecture for Data Stream Management. VLDB Journal 12(2) (2003)
2. Aggarwal, C.C., Yu, P.S.: A Framework for Clustering Uncertain Data Streams. In: IEEE ICDE (2008)
3. Babcock, B., Babu, S., Datar, M., Motwani, R., Widom, J.: Models and Issues in Data Stream Systems. In: ACM PODS (2002)
4. Babu, S., Widom, J.: Continuous Queries over Data Streams. ACM SIGMOD Record 30(3) (2001)
5. Burdick, D., Deshpande, P., Jayram, T.S., Ramakrishnan, R., Vaithyanathan, S.: OLAP over Uncertain and Imprecise Data. In: VLDB (2005)
6. Cai, Y.D., Clutterx, D., Papex, G., Han, J., Welgex, M., Auvilx, L.: MAIDS: Mining Alarming Incidents from Data Streams. In: ACM SIGMOD (2004)
7. Chaudhuri, S., Dayal, U.: An Overview of Data Warehousing and OLAP Technology. ACM SIGMOD Record 26(1) (1997)
8. Chen, Y., Dong, G., Han, J., Wah, B.W., Wang, J.: Multi-Dimensional Regression Analysis of Time-Series Data Streams. In: VLDB (2002)
9. Cormode, G., Garofalakis, M.: Sketching Probabilistic Data Streams. In: ACM SIGMOD (2007)

10. Cormode, G., Korn, F., Tirthapura, S.: Exponentially Decayed Aggregates on Data Streams. In: IEEE ICDE (2008)
11. Cuzzocrea, F., Furfaro, E., Masciari, D.: Improving OLAP Analysis of Multidimensional Data Streams via Efficient Compression Techniques. In: Cuzzocrea, A. (ed.) Intelligent Techniques for Warehousing and Mining Sensor Network Data. IGI Global (2009) (to appear)
12. Cuzzocrea, Wang, W.: Approximate Range-Sum Query Answering on Data Cubes with Probabilistic Guarantees. Journal of Intelligent Information Systems 28(2) (2007)
13. Dalvi, N.N., Suciu, D.: Efficient Query Evaluation on Probabilistic Databases. In: VLDB (2004)
14. Dobra, J., Gehrke, M., Garofalakis, R.: Processing Complex Aggregate Queries over Data Streams. In:ACM SIGMOD (2002)
15. Gaber, M., Zaslavsky, A., Krishnaswamy, S.: Mining Data Streams: A Review. SIGMOD Record 34(2) (2005)
16. Gray, J., Chaudhuri, S., Bosworth, A., Layman, A., Reichart, D., Venkatrao, M., Pellow, F., Pirahesh, H.: Data Cube: A Relational Aggregation Operator Generalizing Group-By, Cross-Tab, and Sub-Totals. Data Mining and Knowledge Discovery 1(1) (1997)
17. Han, J.: OLAP Mining: An Integration of OLAP with Data Mining. IFIP 2.6 DS (1997)
18. Han, J., Chen, Y., Dong, G., Pei, J., Wah, B.W., Wang, J., Cai, Y.D.: Stream Cube: An Architecture for Multi-Dimensional Analysis of Data Streams. Distributed and Parallel Databases 18(2) (2005)
19. Hellerstein, J.M., Haas, P.J., Wang, H.J.: Online Aggregation. In: ACM SIGMOD (1997)
20. Hoeffding, W.: Probability Inequalities for Sums of Bounded Random Variables. Journal of the American Statistical Association 58(301) (1963)
21. Jayram, T.S., McGregor, A., Muthukrishnan, S., Vee, E.: Estimating Statistical Aggregates on Probabilistic Data Streams. In: ACM PODS (2007)
22. Jin, R., Glimcher, L., Jermaine, C., Agrawal, G.: New Sampling-Based Estimators for OLAP Queries. In: IEEE ICDE (2006)
23. Jin, C., Yi, K., Chen, L., Xu Yu, J., Lin, X.: Sliding-Window Top-K Queries on Uncertain Streams. PVLDB 1(1) (2008)
24. Li, X., Han, J., Gonzalez, H.: High-Dimensional OLAP: A Minimal Cubing Approach. In: VLDB (2004)
25. Papoulis: Probability, Random Variables, and Stochastic Processes, 2nd edn. McGraw-Hill, New York (1984)
26. Ré, C., Suciu, D.: Approximate Lineage for Probabilistic Databases. PVLDB 1(1) (2008)
27. Timko, C.E., Dyreson, T.B.: Pre-Aggregation with Probability Distributions. In: ACM DOLAP (2006)
28. Zhang, Q., Li, F., Yi, K.: Finding Frequent Items in Probabilistic Data. In: ACM SIGMOD (2008)

Tractable Probabilistic Description Logic Programs

Thomas Lukasiewicz and Gerardo I. Simari

Abstract. We propose tractable probabilistic description logic programs (dl-programs) for the Semantic Web, which combine tractable description logics (DLs), normal programs under the answer set and the well-founded semantics, and probabilities. In detail, we first provide novel reductions of tight query processing and of deciding consistency in probabilistic dl-programs under the answer set semantics to the answer set semantics of the underlying normal dl-programs. Based on these reductions, we then introduce a novel well-founded semantics for probabilistic dl-programs, called the *total well-founded semantics*. Contrary to the previous answer set and well-founded semantics, it is defined for all probabilistic dl-programs and all probabilistic queries. Furthermore, tight (resp., tight literal) query processing under the total well-founded semantics coincides with (resp., approximates) tight (resp., tight literal) query processing under the previous well-founded (resp., answer set) semantics in all cases where the latter is defined. We then present an anytime algorithm for tight query processing in probabilistic dl-programs under the total well-founded semantics. We also show that tight literal query processing in probabilistic dl-programs under the total well-founded semantics can be done in polynomial time in the data complexity and is complete for EXP in the combined complexity. Finally, we describe an application of probabilistic dl-programs in probabilistic data integration for the Semantic Web.

1 Introduction

During recent years, formalisms for dealing with probabilistic uncertainty have started to play an important role in research related to the Web and the *Semantic Web*, which is an extension of the current Web by standards and technologies that help machines to understand the information on the Web so that they can support richer discovery, data integration, navigation, and automation of tasks [8, 7]. As

Thomas Lukasiewicz · Gerardo I. Simari
Department of Computer Science, University of Oxford
e-mail: {thomas.lukasiewicz,gerardo.simari}@cs.ox.ac.uk

Z. Ma & L. Yan (Eds.): *Advances in Probabilistic Databases*, STUDFUZZ 304, pp. 131–159.
DOI: 10.1007/978-3-642-37509-5_6

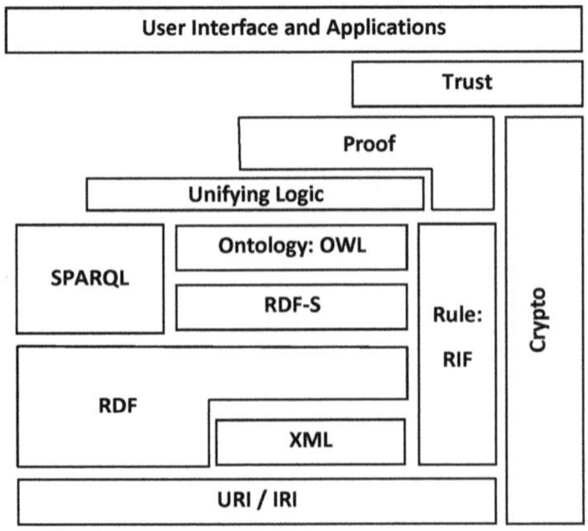

Fig. 1 Hierarchical layers of the Semantic Web

an example of the role of uncertainty in today's Web, note that the order in which Google returns the answers to a Web search query is computed by using probabilistic techniques. Besides Web search and information retrieval, other important Web and Semantic Web applications of formalisms for dealing with probabilistic uncertainty are especially data integration [85] and ontology mapping [67].

The Semantic Web consists of several hierarchical layers, as shown in Fig. 1, where the *Ontology layer*, in the form of the *OWL Web Ontology Language* [86, 38, 87], is currently the highest layer of sufficient maturity. OWL consists of three increasingly expressive sublanguages, namely, *OWL Lite*, *OWL DL*, and *OWL Full*, where OWL Lite and OWL DL are essentially very expressive description logics (DLs) with an RDF syntax [38]. As shown in [37], ontology entailment in OWL Lite (resp., OWL DL) reduces to knowledge base (un)satisfiability in the DL $\mathcal{SHIF}(\mathbf{D})$ (resp., $\mathcal{SHOIN}(\mathbf{D})$). As a next step in the development of the Semantic Web, one currently aims especially at sophisticated reasoning capabilities for the *Rules*, *Logic*, and *Proof layers* of the Semantic Web.

In particular, there is a large body of work on integrating rules and ontologies, which is a key requirement of the layered architecture of the Semantic Web. One type of integration is to build rules on top of ontologies, i.e., for rule-based systems that use vocabulary from ontology knowledge bases. Another form of integration is to build ontologies on top of rules, where ontological definitions are supplemented by rules or imported from rules. Both types of integration have been realized in recent hybrid integrations of rules and ontologies, called *description logic programs* (or *dl-programs*), which are of the form $KB = (L, P)$, where L is a DL knowledge base, and P is a finite set of rules involving queries to L in a loose coupling [25, 26].

Other research efforts are directed towards formalisms for *uncertainty reasoning in the Semantic Web*: An important recent forum for uncertainty in the Semantic Web is the annual *Workshop on Uncertainty Reasoning for the Semantic Web (URSW)* at the *International Semantic Web Conference (ISWC)*; there was also a

W3C Incubator Group on *Uncertainty Reasoning for the World Wide Web*. There are especially extensions of DLs [31], ontology languages [15, 83], and dl-programs [53] by probabilistic uncertainty (to encode ambiguous information, such as "John is a student (resp., teacher) with the probability 0.7 (resp., 0.3)", which is very different from vague/fuzzy information, such as "John is tall with the degree of truth 0.7").

In particular, the probabilistic dl-programs in [53] are one of the most promising approaches to uncertainty reasoning for the Semantic Web, since they faithfully generalize two well-established logic programming and uncertainty formalisms, namely, answer set programming and Bayesian networks, respectively. They also generalize Poole's independent choice logic (ICL) [70], which is a powerful representation and reasoning formalism for single- and also multi-agent systems. The ICL combines logic and probability, and generalizes many important uncertainty formalisms, in particular, influence diagrams, Bayesian networks, Pearl's causal models, Markov decision processes, and normal form games. Moreover, it allows for natural notions of causes and explanations as in Pearl's causal models [27]. It is also closely related to other approaches to probabilistic logic programming, such as P-log [5] and Bayesian logic programs [42].

Since the Web contains a huge amount of data, as an important feature, Web and Semantic Web formalisms should allow for efficient algorithms. However, no such algorithms were known so far for the probabilistic dl-programs in [53]. In this work, we aim at filling this gap. We propose an approach to probabilistic dl-programs that is defined on top of tractable DLs (rather than $\mathcal{SHIF}(\mathbf{D})$ and $\mathcal{SHOIN}(\mathbf{D})$ as in [53]), and show that this approach allows for tight query processing with polynomial data complexity. In the course of this, we also provide some other new results around probabilistic dl-programs, which are briefly summarized as follows:

- We provide novel reductions of deciding consistency and of tight query processing in probabilistic dl-programs under the answer set semantics to computing the answer sets of the underlying normal dl-programs. These reductions significantly simplify previous reductions proposed in [53], which additionally require to decide the solvability of a (in general quite large) system of linear inequalities and to solve two linear optimization problems relative to them, respectively.
- We define a novel well-founded semantics of probabilistic dl-programs, called the *total well-founded semantics*, since it defines tight answers for *all* probabilistic queries. This is in contrast to the previous well-founded semantics of probabilistic dl-programs in [53], which defines tight answers only for a quite restricted class of probabilistic queries. The total well-founded semantics is also defined for all probabilistic dl-programs, contrary to the answer set semantics, which is only defined for consistent probabilistic dl-programs.
- As for other nice semantic features of the total well-founded semantics, the tight answers under the total well-founded semantics coincide with the tight answers under the well-founded semantics of [53], if the latter are defined. For literal queries, the tight answers under the total well-founded semantics approximate the tight answers under the answer set semantics, if the latter are defined.

- We provide an anytime algorithm for tight query processing in probabilistic dl-programs under the total well-founded semantics, along with a precise characterization of its anytime error. Furthermore, we show that tight query processing under the total well-founded semantics can be done in polynomial time in the data complexity and is complete for EXP in the combined complexity.
- We describe an application of probabilistic dl-programs in probabilistic data integration for the Semantic Web, where probabilistic dl-programs allow for dealing with probabilistic uncertainty and inconsistencies. We especially discuss different types of probabilistic data integration that can be realized with our approach.

The rest of this paper is organized as follows. In Sections 2 and 3, we recall tractable DLs and dl-programs under the answer set and the well-founded semantics, respectively. Section 4 recalls probabilistic dl-programs. In Section 5, we provide a new reduction of tight query processing in probabilistic dl-programs under the answer set semantics to the answer set semantics of normal dl-programs. Section 6 introduces the total well-founded semantics of probabilistic dl-programs. In Section 7, we provide an anytime algorithm for tight query processing in probabilistic dl-programs under the total well-founded semantics, as well as tractability and complexity results. Section 8 describes an application of probabilistic dl-programs in probabilistic data integration for the Semantic Web. In Section 9, we discuss related work. Section 10 summarizes our results, and gives an outlook on future research.

2 Description Logics

The probabilistic dl-programs of this paper assume that the underlying description logic (DL) allows for decidable conjunctive query processing. The tractability and complexity results (see Section 7.2) also assume that the underlying DL allows for conjunctive query processing in polynomial data complexity. We use *DL-Lite* here, but the tractability and complexity results also hold for the variants of *DL-Lite* in [11, 68]. We now recall the syntax and the semantics of *DL-Lite*. Intuitively, DLs model a domain of interest in terms of concepts and roles, which represent classes of individuals and binary relations between classes of individuals, respectively.

2.1 Syntax

We first define concepts and axioms, and then knowledge bases and conjunctive queries in *DL-Lite*. We assume pairwise disjoint sets $\mathbf{A}$, $\mathbf{R}$, and $\mathbf{I}$ of *atomic concepts*, *(atomic) roles*, and *individuals*, respectively. We use $\mathbf{R}^-$ to denote the set of all inverses R^- of roles $R \in \mathbf{R}$. A *basic concept* B is either an atomic concept $A \in \mathbf{A}$ or an *existential role restriction* $\exists R$, where $R \in \mathbf{R} \cup \mathbf{R}^-$. An *axiom* is either:

- A *concept inclusion axiom* $B \sqsubseteq C$, where B is a basic concept, and C is either a basic concept B' or its negation $\neg B'$ (called *concept*),
- a *concept membership axiom* $B(x)$, where B is a basic concept and $x \in \mathbf{I}$,

- a *role membership axiom* $R(x,y)$, where $R \in \mathbf{R}$ and $x,y \in \mathbf{I}$,
- a *functionality axiom* $(\mathsf{funct}\, R)$, where $R \in \mathbf{R} \cup \mathbf{R}^-$.

Given the basic definitions above, a *(DL) knowledge base* L is a finite set of axioms; a *conjunctive query* over L is of the form

$$Q(\mathbf{x}) = \exists \mathbf{y}\, (conj(\mathbf{x},\mathbf{y})), \tag{1}$$

where $\mathbf{x}$ and $\mathbf{y}$ are tuples of distinct variables, and $conj(\mathbf{x},\mathbf{y})$ is a conjunction of assertions $B(z)$ and $R(z_1,z_2)$, where B and R are basic concepts and roles from $\mathbf{R}$, respectively, and z, z_1, and z_2 are individuals from $\mathbf{I}$ or variables in $\mathbf{x}$ or $\mathbf{y}$.

Example 1. An online store (such as *amazon.com*) may use a DL knowledge base to classify and characterize its products. For example, suppose that (1) textbooks are books, (2) personal computers and cameras are electronic products, (3) books and electronic products are products, (4) every product has at least one related product, (5) only products are related to each other, (6) *tb_ai* and *tb_lp* are textbooks, which are related to each other, (7) *pc_ibm* and *pc_hp* are personal computers, which are related to each other, and (8) *ibm* and *hp* are providers for *pc_ibm* and *pc_hp*, respectively. This knowledge is expressed by the following *DL-Lite* knowledge base L:

(1) $\mathit{Textbook} \sqsubseteq \mathit{Book}$;
(2) $\mathit{PC} \sqcup \mathit{Camera} \sqsubseteq \mathit{Electronics}$;
(3) $\mathit{Book} \sqcup \mathit{Electronics} \sqsubseteq \mathit{Product}$;
(4) $\mathit{Product} \sqsubseteq \exists \mathit{related}$;
(5) $\mathit{related} \sqcup \mathit{related}^- \sqsubseteq \mathit{Product}$;
(6) $\mathit{Textbook}(tb\_ai)$ $\mathit{Textbook}(tb\_lp)$; $\mathit{related}(tb\_ai, tb\_lp)$;
(7) $\mathit{PC}(pc\_ibm)$; $\mathit{PC}(pc\_hp)$; $\mathit{related}(pc\_ibm, pc\_hp)$;
(8) $\mathit{provides}(ibm, pc\_ibm)$; $\mathit{provides}(hp, pc\_hp)$.

2.2 Semantics

The semantics of *DL-Lite* is defined as usual in first-order logics. An *interpretation* $\mathcal{I} = (\Delta^{\mathcal{I}}, \cdot^{\mathcal{I}})$ consists of a nonempty *domain* $\Delta^{\mathcal{I}}$ and a mapping $\cdot^{\mathcal{I}}$ that assigns to each $A \in \mathbf{A}$ a subset of $\Delta^{\mathcal{I}}$, to each $o \in \mathbf{I}$ an element of $\Delta^{\mathcal{I}}$ (such that $o_1 \neq o_2$ implies $o_1^{\mathcal{I}} \neq o_2^{\mathcal{I}}$, commonly referred to as the *unique name assumption*), and to each $R \in \mathbf{R}$ a subset of $\Delta^{\mathcal{I}} \times \Delta^{\mathcal{I}}$. We extend $\cdot^{\mathcal{I}}$ to all concepts and roles as follows:

- $(\neg B)^{\mathcal{I}} = \Delta^{\mathcal{I}} \setminus B^{\mathcal{I}}$, for all basic concepts B;
- $(\exists R)^{\mathcal{I}} = \{x \mid \exists y\colon (x,y) \in R^{\mathcal{I}}\}$, for all roles $R \in \mathbf{R} \cup \mathbf{R}^-$;
- $(R^-)^{\mathcal{I}} = \{(y,x) \mid (x,y) \in R^{\mathcal{I}}\}$, for all atomic roles $R \in \mathbf{R}$.

Next, we define the *satisfaction* of an axiom F by $\mathcal{I}$, denoted $\mathcal{I} \models F$, as usual:

- $\mathcal{I} \models B \sqsubseteq C$ iff $B^{\mathcal{I}} \subseteq C^{\mathcal{I}}$, for all basic concepts B and concepts C;
- $\mathcal{I} \models B(a)$ iff $a^{\mathcal{I}} \in B^{\mathcal{I}}$, for all basic concepts B;
- $\mathcal{I} \models R(a,b)$ iff $(a^{\mathcal{I}}, b^{\mathcal{I}}) \in R^{\mathcal{I}}$, for all atomic roles $R \in \mathbf{R}$;
- $\mathcal{I} \models (\mathsf{funct}\, R)$ iff $(x,y) \in R^{\mathcal{I}} \wedge (x,z) \in R^{\mathcal{I}} \Rightarrow y = z$, for all roles $R \in \mathbf{R} \cup \mathbf{R}^-$.

An interpretation $\mathcal{I}$ *satisfies* the axiom F, or $\mathcal{I}$ is a *model* of F, iff $\mathcal{I} \models F$. Furthermore, $\mathcal{I}$ *satisfies* a knowledge base L, or $\mathcal{I}$ is a *model* of L, denoted $\mathcal{I} \models L$, iff $\mathcal{I} \models F$ for all $F \in L$. We say that L is *satisfiable* (resp., *unsatisfiable*) iff L has a (resp., no) model. An axiom F is a *logical consequence* of L, denoted $L \models F$, iff every model of L satisfies F. A negated axiom $\neg F$ is a *logical consequence* of L, denoted $L \models \neg F$, iff every model of L does not satisfy F.

A tuple $\mathbf{c}$ of individuals from $\mathbf{I}$ is an *answer* for a conjunctive query $Q(\mathbf{x}) = \exists \mathbf{y}\,(conj(\mathbf{x},\mathbf{y}))$ to a knowledge base L iff for every model $\mathcal{I} = (\Delta^{\mathcal{I}}, \cdot^{\mathcal{I}})$ of L, there exists a tuple $\mathbf{o}$ of elements from $\Delta^{\mathcal{I}}$ such that all assertions in $conj(\mathbf{c},\mathbf{o})$ are satisfied in $\mathcal{I}$. In *DL-Lite*, computing all such answers has a polynomial data complexity.

3 Description Logic Programs

We adopt the description logic programs (or dl-programs) of [25, 26], which consist of a DL knowledge base L and a generalized normal program P, which may contain queries to L (called dl-queries). Note that these dl-programs can be extended by queries to other formalisms, such as RDF theories [24]. We first define the syntax of dl-programs and then their answer set and their well-founded semantics. Note that in contrast to [25, 26], we assume here that dl-queries may be conjunctive queries to L.

3.1 Syntax

We assume a function-free first-order vocabulary Φ with finite nonempty sets of constant and predicate symbols Φ_c and Φ_p, respectively, and a set of variables $\mathcal{X}$. We make the following assumptions:

- Φ_c is a subset of $\mathbf{I}$ (since the constants in Φ_c may occur in concept and role assertions of dl-queries); and
- Φ and $\mathbf{A}$ (resp., $\mathbf{R}$) have no unary (resp., binary) predicate symbols in common (and thus dl-queries are the only interface between L and P).

A *term* is a constant symbol from Φ or a variable from $\mathcal{X}$. If p is a predicate symbol of arity $k \geqslant 0$ from Φ, and $t_1,\ldots,t_k$ are terms, then $p(t_1,\ldots,t_k)$ is an *atom*. A *literal* is an atom a or a default-negated atom $not\, a$. A *(normal) rule* r is of the form

$$a \leftarrow b_1,\ldots,b_k, not\ b_{k+1},\ldots, not\ b_m\,, \tag{2}$$

where $a, b_1,\ldots,b_m$ are atoms and $m \geqslant k \geqslant 0$. We call a the *head* of r, denoted $H(r)$, while the conjunction $b_1,\ldots,b_k, not\ b_{k+1},\ldots, not\ b_m$ is the *body* of r; its *positive* (resp., *negative*) part is $b_1,\ldots,b_k$ (resp., $not\ b_{k+1},\ldots, not\ b_m$). We define $B(r)$ as the union of $B^+(r) = \{b_1,\ldots,b_k\}$ and $B^-(r) = \{b_{k+1},\ldots,b_m\}$. A *(normal) program* P is a finite set of (normal) rules. We say P is *positive* iff it is "*not*"-free.

A *dl-query* $Q(\mathbf{t})$ is a conjunctive query of the form (1). A *dl-atom* has the form

$$DL[S_1 \uplus p_1,\ldots,S_m \uplus p_m; Q(\mathbf{t})]\,,$$

where each S_i is a concept or role, p_i is a unary resp. binary predicate symbol, $Q(\mathbf{t})$ is a dl-query, and $m \geqslant 0$. We call $p_1,\ldots,p_m$ its *input predicate symbols*. Intuitively,

$\uplus$ increases S_i by the extension of p_i. A *(normal) dl-rule* r is of the form (2), where any $b \in B(r)$ may be a dl-atom. A *(normal) dl-program* $KB = (L, P)$ consists of a DL knowledge base L and a finite set of (normal) dl-rules P. We say $KB = (L, P)$ is *positive* iff P is positive. *Ground terms, atoms, literals*, etc., are defined as usual. We denote by $ground(P)$ the set of all ground instances of dl-rules in P relative to Φ_c.

Example 2. Consider the dl-program $KB = (L, P)$, where L is the DL knowledge base from Example 1, and P is the following set of dl-rules:

(1) $pc(pc\_1);\ \ pc(pc\_2);\ \ pc(pc\_3);$
(2) $brand\_new(pc\_1);\ \ brand\_new(pc\_2);$
(3) $vendor(dell, pc\_1);\ \ vendor(dell, pc\_2);\ \ vendor(dell, pc\_3);$
(4) $avoid(X) \leftarrow DL[PC \uplus pc; PC](X), not\ offer(X);$
(5) $offer(X) \leftarrow DL[PC \uplus pc; Electronics](X), not\ brand\_new(X);$
(6) $provider(V) \leftarrow vendor(V, X), DL[PC \uplus pc; Product](X);$
(7) $provider(V) \leftarrow DL[provides](V, X), DL[PC \uplus pc; Product](X);$
(8) $similar(X, Y) \leftarrow DL[related](X, Y);$
(9) $similar(X, Y) \leftarrow pc(X), pc(Y), X \neq Y;$
(10) $similar(X, Z) \leftarrow similar(X, Y), similar(Y, Z), X \neq Z;$
(11) $buyPC(X) \leftarrow DL[PC \uplus pc; PC](X), not\ avoid(X), not\ exclude(X);$
(12) $exclude(X) \leftarrow DL[PC \uplus pc; PC](X), buyPC(Y), similar(X, Y).$

The above dl-rules express that (1) *pc_1*, *pc_2*, and *pc_3* are additional personal computers, (2) *pc_1* and *pc_2* are brand new, (3) *dell* is the vendor of *pc_1*, *pc_2*, and *pc_3*, (4) a customer avoids all PCs that are not on offer, (5) all electronic products that are not brand new are on offer, (6) every vendor of a product is a provider, (7) every entity providing a product is a provider, (8) all related products are similar, (9) all PCs are similar to each other (but a PC is not similar to itself), (10) the binary similarity relation on products is transitively closed, and (11), (12) a customer buys a PC if he does not avoid it and has not decided to buy a similar one already.

3.2 Answer Set Semantics

The *Herbrand base* HB_Φ is the set of all ground atoms constructed from constant and predicate symbols in Φ. An *interpretation* I is any $I \subseteq HB_\Phi$. We say I is a *model* of $a \in HB_\Phi$ under a DL knowledge base L, denoted $I \models_L a$, iff $a \in I$. We say I is a *model* of a ground dl-atom $a = DL[S_1 \uplus p_1, \ldots, S_m \uplus p_m; Q(\mathbf{c})]$ under L, denoted $I \models_L a$, iff $L \cup \bigcup_{i=1}^{m} A_i(I) \models Q(\mathbf{c})$, where $A_i(I) = \{S_i(\mathbf{e}) \mid p_i(\mathbf{e}) \in I\}$. We say I is a *model* of a ground dl-rule r iff $I \models_L H(r)$ whenever $I \models_L B(r)$, i.e., $I \models_L a$ for all $a \in B^+(r)$ and $I \not\models_L a$ for all $a \in B^-(r)$. We say I is a *model* of a dl-program $KB = (L, P)$, denoted $I \models KB$, iff $I \models_L r$ for all $r \in ground(P)$.

Like ordinary positive programs, each positive dl-program KB has a unique least model, denoted M_{KB}, which naturally characterizes its semantics. The *answer set semantics* of general dl-programs is then defined by a reduction to the least model semantics of positive ones, using a reduct that generalizes the ordinary Gelfond-Lifschitz reduct [30] and removes all default-negated atoms in dl-rules: For dl-programs $KB = (L, P)$, the *dl-reduct* of P relative to L and an interpretation $I \subseteq HB_\Phi$, denoted P^I_L, is the set of all dl-rules obtained from $ground(P)$ by:

- deleting each dl-rule r such that $I \models_L a$ for some $a \in B^-(r)$, and
- deleting from each remaining dl-rule r the negative body.

An *answer set* of KB is an interpretation $I \subseteq HB_\Phi$ such that I is the unique least model of (L, P^I_L). A dl-program is *consistent* iff it has an answer set.

Example 3. Consider the dl-program $KB = (L, P)$ from Example 2, which in turn relies on the *DL-Lite* knowledge base L from Example 1. Intuitively, any answer set of KB contains the atoms $\mathit{offer}(pc_3)$, $\mathit{offer}(pc\_ibm)$, and $\mathit{offer}(pc\_hp)$ (since none of these PCs is brand new), $\mathit{avoid}(pc_1)$ and $\mathit{avoid}(pc_2)$ (since they are not on offer), $\mathit{provider}(dell)$, and a set of atoms for *similar* consisting of all possible irreflexive pairs of objects in the set $\{pc\_1, pc\_2, pc\_3, pc\_ibm, pc\_hp\}$.

On the other hand, all answer sets will differ with respect to the *buyPC* atom. Rule (11) in Example 2 states that $\mathit{buyPC}(X)$ is true only if X corresponds to a PC (via the query to the *DL-Lite* knowledge base, augmented with the *pc* predicate), there is no $\mathit{avoid}(X)$ atom in the answer set, and there is no other $\mathit{buyPC}(Y)$ atom in the knowledge base, where Y is similar to X. Therefore, there will be three answer sets, each containing one of $\mathit{buyPC}(pc_3)$, $\mathit{buyPC}(pc\_ibm)$, and $\mathit{buyPC}(pc\_hp)$, as well as $\mathit{exclude}(X)$ atoms for the other two objects.

The answer set semantics of dl-programs has several nice features. In particular, for dl-programs $KB = (L, P)$ without dl-atoms, it coincides with the ordinary answer set semantics of P. Answer sets of a general dl-program KB are also minimal models of KB. Furthermore, positive and locally stratified dl-programs have exactly one answer set, which coincides with their canonical minimal model.

3.3 Well-Founded Semantics

Rather than associating with every dl-program a (possibly empty) set of two-valued interpretations, the well-founded semantics associates with every dl-program a unique three-valued interpretation.

A *(classical) literal* is either an atom a or its negation $\neg a$. For sets $S \subseteq HB_\Phi$, we define $\neg S = \{\neg a \mid a \in S\}$. We define $Lit_\Phi = HB_\Phi \cup \neg HB_\Phi$. A set of ground classical literals $S \subseteq Lit_\Phi$ is *consistent* iff $S \cap \{a, \neg a\} = \emptyset$ for all $a \in HB_\Phi$. A *three-valued interpretation* is any consistent $I \subseteq Lit_\Phi$. We define the well-founded semantics of dl-programs $KB = (L, P)$ via a generalization of the operator γ^2 for ordinary normal programs. We define the operator γ_{KB} as follows. For every $I \subseteq HB_\Phi$, we define $\gamma_{KB}(I)$ as the least model of the positive dl-program $KB^I = (L, P^I_L)$. The operator γ_{KB} is anti-monotonic, and thus the operator γ^2_{KB} (defined by $\gamma^2_{KB}(I) = \gamma_{KB}(\gamma_{KB}(I))$, for every $I \subseteq HB_\Phi$) is monotonic and has a least and a greatest fixpoint, denoted $\mathit{lfp}(\gamma^2_{KB})$ and $\mathit{gfp}(\gamma^2_{KB})$, respectively. Then, the *well-founded semantics* of the dl-program KB, denoted $\mathit{WFS}(KB)$, is defined as $\mathit{lfp}(\gamma^2_{KB}) \cup \neg(HB_\Phi - \mathit{gfp}(\gamma^2_{KB}))$.

Example 4. Consider once again the dl-program $KB = (L, P)$ from Example 2, where L is as in Example 1. The well-founded semantics of KB contains all the atoms in all answer sets in Example 3, but no atoms among $buyPC(pc_3)$, $buyPC(pc\_ibm)$, $buyPC(pc\_hp)$, $exclude(pc_3)$, $exclude(pc\_ibm)$, and $exclude(pc\_hp)$ (and their negations); such atoms are thus *undefined* under the well-founded semantics of KB.

As an important property, the well-founded semantics for dl-programs approximates their answer set semantics. That is, for all consistent dl-programs KB and literals $\ell \in Lit_{\Phi}$, every $\ell \in WFS(KB)$ is true in every answer set of KB.

4 Probabilistic Description Logic Programs

In this section, we recall probabilistic dl-programs from [53], which are defined as a combination of dl-programs with Poole's independent choice logic (ICL)[70]. The ICL is based on ordinary acyclic logic programs under different "choices", where every choice along with an acyclic logic program produces a first-order model, and one then obtains a probability distribution over the set of all first-order models by placing a probability distribution over the different choices. Informally, differently from the ICL, probabilistic dl-programs consist of a dl-program (L, P) and a probability distribution μ over a set of total choices B. Every total choice B along with the dl-program (L, P) then defines a set of Herbrand interpretations of which the probabilities sum up to $\mu(B)$. We now first define the syntax of probabilistic dl-programs and probabilistic queries, and then their answer set semantics.

4.1 Syntax

We now define the syntax of probabilistic dl-programs and probabilistic queries addressed to them. We first define choice spaces and probabilities on choice spaces.

We assume a function-free first-order vocabulary Φ with nonempty finite sets of constant and predicate symbols, and a set of variables $\mathcal{X}$, as above for dl-programs. We use HB_{Φ} (resp., HU_{Φ}) to denote the Herbrand base (resp., universe) over Φ. In the sequel, we assume that HB_{Φ} is nonempty. An *event* α is any Boolean combination of atoms (i.e., constructed from atoms via the Boolean operators "$\wedge$" and "$\neg$"). A *conditional event* is of the form $\beta|\alpha$, where α and β are events. Ground terms, ground events, and substitutions are defined as usual.

A *choice space* C is a set of pairwise disjoint and nonempty sets $A \subseteq HB_{\Phi}$. Any $A \in C$ is called an *alternative* of C, and any element $a \in A$ is called an *atomic choice* of C. Intuitively, every alternative $A \in C$ represents a random variable and every atomic choice $a \in A$ one of its possible values. A *total choice* of C is a set $B \subseteq HB_{\Phi}$ such that $|B \cap A| = 1$ for all $A \in C$ (and thus $|B| = |C|$). Intuitively, every total choice B of C represents an assignment of values to all the random variables. A *probability* μ on a choice space C is a probability function on the set of all total choices of C. Intuitively, every probability μ is a probability distribution over the set of all variable assignments. Since C and all its alternatives are finite, μ can be defined by

- a mapping $\mu\colon \bigcup C \to [0,1]$ such that $\sum_{a \in A} \mu(a) = 1$ for all $A \in C$, and
- $\mu(B) = \Pi_{b \in B} \mu(b)$ for all total choices B of C.

Intuitively, the first condition defines a probability over the values of each random variable of C, and the second assumes independence between the random variables.

A *probabilistic dl-program* $KB = (L, P, C, \mu)$ consists of

- a dl-program (L, P),
- a choice space C such that (i) $\bigcup C \subseteq HB_\Phi$ and (ii) no atomic choice in C coincides with the head of any $r \in ground(P)$, and
- a probability μ on C.

Intuitively, since the total choices of C select subsets of P, and μ is a probability distribution on the total choices of C, every probabilistic dl-program is the compact encoding of a probability distribution on a finite set of normal dl-programs. Observe here that P is fully general and not necessarily stratified or acyclic. A *probabilistic query* has the form

$$\exists(\beta | \alpha)[r, s],$$

where $\beta | \alpha$ is a conditional event, and r and s are variables.

Example 5. Consider the probabilistic dl-program $KB = (L, P, C, \mu)$, where L and P are as in Examples 1 and 2, except that the dl-rules (4), (5), and (11) are replaced by the dl-rules (4′), (5′), and (11′), respectively, and the dl-rule (13) is added:

$$\begin{array}{ll}
(4') & avoid(X) \leftarrow DL[PC \uplus pc; Electronics](X), not\ offer(X), avoid_pos;\\
(5') & offer(X) \leftarrow DL[PC \uplus pc; Electronics](X), not\ brand_new(X), offer_pos;\\
(11') & buyPC(X) \leftarrow DL[PC \uplus pc; PC](X), not\ avoid(X), not\ exclude(X), one_buy_pos;\\
(13) & buyAccessory(X) \leftarrow DL[Electronics](X), not\ avoid(X), buyPC(Y),\\
 & \qquad not\ DL[PC \uplus pc; PC](X), acc_buy_pos.
\end{array}$$

Let C be defined as

$$\{\{avoid\_pos, avoid\_neg\}, \{offer\_pos, offer\_neg\},\\ \{one\_buy\_pos, one\_buy\_neg\}, \{acc\_buy\_pos, acc\_buy\_neg\}\},$$

and let μ be given as follows:

- $\mu(avoid\_pos) = 0.7$, $\mu(avoid\_neg) = 0.3$,
- $\mu(offer\_pos) = 0.7$, $\mu(offer\_neg) = 0.3$,
- $\mu(one\_buy\_pos) = 0.95$, $\mu(one\_buy\_neg) = 0.05$,
- $\mu(acc\_buy\_pos) = 0.8$, $\mu(acc\_buy\_neg) = 0.2$.

The new dl-rules (4′) and (5′) express that the original dl-rules (4) and (5) now only hold with probability 0.7. Furthermore, (11′) expresses that a customer buys a single PC with probability 0.95, while rule (13) states that a customer buying a PC may choose to also buy a non-PC electronics item as an accessory with probability 0.8.

In a probabilistic query, one may ask for the tight probability bounds that a customer buys a webcam w, if (i) PC p is bought, (ii) p is on offer, and (iii) w is not on offer; the result to this query may, for instance, help to decide whether it is useful to automatically show a product w to a customer buying the product p):

$$\exists(buyAccessory(w) \,|\, \neg offer(w) \wedge buyPC(p) \wedge offer(p))[R, S].$$

4.2 Answer Set Semantics

We next define an answer set semantics of probabilistic dl-programs, the consistency of such programs, and tight answers to probabilistic queries.

Given a probabilistic dl-program $KB = (L, P, C, \mu)$, a *probabilistic interpretation* Pr is a probability function on the set of all $I \subseteq HB_\Phi$. We say that Pr is an *answer set* of KB iff the following two conditions hold:

- every interpretation $I \subseteq HB_\Phi$ with $Pr(I) > 0$ is an answer set of the dl-program $(L, P \cup \{p \leftarrow \mid p \in B\})$ for some total choice B of C (which implies $B \subseteq I$), and
- $Pr\left(\bigwedge_{p \in B} p\right) = \mu(B)$ for every total choice B of C.

Informally, these conditions state that Pr is an answer set of $KB = (L, P, C, \mu)$ iff (i) every interpretation $I \subseteq HB_\Phi$ of positive probability under Pr is an answer set of the dl-program (L, P) under some total choice B of C, and (ii) Pr coincides with μ on the total choices B of C. We say KB is *consistent* iff it has an answer set Pr.

Given a ground event α, the *probability* of α in a probabilistic interpretation Pr, denoted $Pr(\alpha)$, is the sum of all $Pr(I)$ such that $I \subseteq HB_\Phi$ and $I \models \alpha$. We say that $(\beta|\alpha)[l, u]$ (where $l, u \in [0, 1]$) is a *tight consequence* of a consistent probabilistic dl-program KB under the answer set semantics iff l (resp., u) is the infimum (resp., supremum) of $Pr(\alpha \wedge \beta) / Pr(\alpha)$ subject to all answer sets Pr of KB with $Pr(\alpha) > 0$ (note that this infimum (resp., supremum) is naturally defined as 1 (resp., 0) iff no such Pr exists). The *tight answer* for a probabilistic query $Q = \exists(\beta|\alpha)[r, s]$ to KB under the answer set semantics is the set of all ground substitutions θ (for the variables in Q) such that $(\beta|\alpha)[r, s]\theta$ is a tight consequence of KB under the answer set semantics. For ease of presentation, since the tight answers for probabilistic queries of the form $Q = \exists(\beta|\alpha)[r, s]$ with non-ground $\beta|\alpha$ can be reduced to the tight answers for probabilistic queries $Q = \exists(\beta|\alpha)[r, s]$ with ground $\beta|\alpha$, we consider only the latter type of probabilistic queries from now on.

5 Novel Answer Set Characterizations

In this section, we give novel characterizations of (i) the consistency of probabilistic dl-programs and (ii) tight query processing in consistent probabilistic dl-programs under the answer set semantics in terms of the answer sets of normal dl-programs.

As shown in [53], a probabilistic dl-program $KB = (L, P, C, \mu)$ is consistent iff the system of linear constraints $LC_\top$ (see Fig. 2) over y_r ($r \in R$) is solvable. Here, R is the union of all sets of answer sets of $(L, P \cup \{p \leftarrow \mid p \in B\})$ for all total choices B of C. Observe, however, that $LC_\top$ is defined over a set of variables R that corresponds to the set of all answer sets of the underlying normal dl-programs, and thus R is in general quite large.

Example 6. Consider the probabilistic dl-program $KB = (L, P, C, \mu)$, where L and P are as in Examples 1 and 2, except that the dl-rules (4) and (11) are replaced by the following dl-rules (4′) and (11′), respectively:

(4′) $avoid(X) \leftarrow DL[PC \uplus pc; PC](X), not\ offer(X), avoid\_pos$;
(11′) $buyPC(X) \leftarrow DL[PC \uplus pc; PC](X), not\ avoid(X), not\ exclude(X), buy\_pos$.

$$\begin{aligned}
\sum_{r\in R,\, r\not\models\bigwedge B} -\mu(B)\, y_r + \sum_{r\in R,\, r\models\bigwedge B} (1-\mu(B))\, y_r &= 0 \quad \text{(for all total choices } B \text{ of } C\text{)}\\
\sum_{r\in R,\, r\models\alpha} y_r &= 1\\
y_r &\geqslant 0 \quad \text{(for all } r\in R\text{)}
\end{aligned}$$

Fig. 2 System of linear constraints LC_α

The choice space C is defined as

$$\{\{avoid\_pos, avoid\_neg\}, \{buy\_pos, buy\_neg\}\},$$

and μ is given as follows:

- $\mu(avoid\_pos) = 0.5$, $\mu(avoid\_neg) = 0.5$,
- $\mu(buy\_pos) = 0.8$, $\mu(buy\_neg) = 0.2$.

We then obtain four total choices:

$$\{avoid\_pos, buy\_pos\},\ \{avoid\_pos, buy\_neg\},$$
$$\{avoid\_neg, buy\_pos\},\ \{avoid\_neg, buy\_neg\}.$$

The set of linear constraints $LC_\top$ for *KB* then comprises four constraints corresponding to each of the total choices, one constraint corresponding to the sum to 1, and one constraint for each variable, expressing its non-negativeness. There is one variable for each possible answer set given all possible total choices. Intuitively, we can see that multiple answer sets can arise only if *buy_pos* is true. The case where *avoid_pos* is also true was investigated in Example 3, where we saw that there are three answer sets in this situation. The remaining case corresponds to *avoid_neg* being true; here, rule (4′) no longer applies, and therefore rule (11′)'s *not avoid*(X) atom will hold for all answers to the query in the dl-atom $DL[PC \uplus pc; PC](X)$, which contains five objects: pc_1, pc_2, pc_3, *pc_hp*, and *pc_ibm*. Therefore, we have five more answer sets in this case, and thus a total of 10 variables in $LC_\top$.

The following theorem shows that the consistency of probabilistic dl-programs can be expressed in terms of answer sets of normal dl-programs only, without having to additionally decide whether or not a system of linear constraints is solvable.

Theorem 1 (Consistency). *Let* $KB = (L, P, C, \mu)$ *be a probabilistic dl-program. Then, KB is consistent iff, for every total choice B of C such that* $\mu(B) > 0$, *the dl-program* $(L, P \cup \{p \leftarrow \mid p \in B\})$ *is consistent.*

Similarly, as shown in [53], computing tight answers for probabilistic queries can be reduced to computing all answer sets of normal dl-programs and solving two linear optimization problems. More specifically, let $KB = (L, P, C, \mu)$ be a consistent probabilistic dl-program, and let $Q = \exists(\beta|\alpha)[r, s]$ be a probabilistic query with ground $\beta|\alpha$. Then, the tight answer for Q to *KB* is given by $\theta = \{r/l,\ s/u\}$, where l (resp.,

u) is the optimal value of the subsequent linear program (3) over y_r ($r \in R$), if (3) has a solution, and it is given by $\theta = \{r/1, s/0\}$, if (3) has no solution.

$$\min \text{ (resp., max) } \textstyle\sum_{r \in R, r \models \alpha \wedge \beta} y_r \quad \text{subject to } LC_\alpha \text{ (see Fig. 2).} \tag{3}$$

But the linear program (3) is defined over the same (generally quite large) set of variables as the system of linear constraints $LC_\top$ above. The following theorem shows that the tight answers can also be expressed in terms of answer sets of normal dl-programs only, without additionally solving two linear optimization problems.

Theorem 2 (Tight Query Processing). *Let $KB = (L, P, C, \mu)$ be a consistent probabilistic dl-program, and let $Q = \exists(\beta|\alpha)[r,s]$ be a probabilistic query with ground $\beta|\alpha$. Let a (resp., b) be the sum of all $\mu(B)$ such that (i) B is a total choice of C and (ii) $\alpha \wedge \beta$ is true in every (resp., some) answer set of $(L, P \cup \{p \leftarrow \mid p \in B\})$. Let c (resp., d) be the sum of all $\mu(B)$ such that (i) B is a total choice of C and (ii) $\alpha \wedge \neg\beta$ is true in every (resp., some) answer set of $(L, P \cup \{p \leftarrow \mid p \in B\})$. Then, the tight answer θ for Q to KB under the answer set semantics is given as follows:*

$$\theta = \begin{cases} \{r/1, s/0\} & \text{if } b=0 \text{ and } d=0; \\ \{r/0, s/0\} & \text{if } b=0 \text{ and } d \neq 0; \\ \{r/1, s/1\} & \text{if } b \neq 0 \text{ and } d=0; \\ \{r/\frac{a}{a+d}, s/\frac{b}{b+c}\} & \text{otherwise.} \end{cases}$$

6 Total Well-Founded Semantics

In this section, we define a novel well-founded semantics for probabilistic dl-programs, called the *total well-founded semantics*, since it is defined for all probabilistic queries to probabilistic dl-programs, as opposed to the well-founded semantics of [53], which is only defined for a very limited class of probabilistic queries. Furthermore, the total well-founded semantics is defined for all probabilistic dl-programs, as opposed to the answer set semantics, which is only defined for consistent ones.

More concretely, given a probabilistic dl-program $KB = (L, P, C, \mu)$ and a probabilistic query $Q = \exists(\beta|\alpha)[r,s]$ with ground $\beta|\alpha$, the tight answer θ for Q to KB under the well-founded semantics of [53] *exists* iff both ground events $\alpha \wedge \beta$ and α are defined in every $S = WFS(L, P \cup \{p \leftarrow | p \in B\})$ such that B is a total choice of C. Here, a ground event ϕ is *defined* in S iff either $I \models \phi$ for every interpretation $I \subseteq HB_\Phi$ such that (i) $S \cap HB_\Phi \subseteq I$ and (ii) $\neg S \cap I = \emptyset$, or $I \not\models \phi$ for every interpretation $I \subseteq HB_\Phi$ such that (i) $S \cap HB_\Phi \subseteq I$ and (ii) $\neg S \cap I = \emptyset$. If α is false in every $WFS(L, P \cup \{p \leftarrow | p \in B\})$ such that B is a total choice of C, then the tight answer is defined as $\theta = \{r/1, s/0\}$; otherwise, the tight answer (if it exists) is defined as $\theta = \{r/\frac{u}{v}, s/\frac{u}{v}\}$, where u (resp., v) is the sum of all $\mu(B)$ such that:

- B is a total choice of C, and
- $\alpha \wedge \beta$ (resp., α) is true in $WFS(L, P \cup \{p \leftarrow \mid p \in B\})$.

We define the total well-founded semantics as follows, taking inspiration from the novel answer set characterization of tight answers in the previous section.

Definition 1 (Total Well-Founded Semantics). Let $KB = (L, P, C, \mu)$ be a probabilistic dl-program, and let $Q = \exists(\beta|\alpha)[r,s]$ be a probabilistic query with ground $\beta|\alpha$. Let a (resp., b^-) be the sum of all $\mu(B)$ such that (i) B is a total choice of C and (ii) $\alpha \wedge \beta$ is true (resp., false) in $WFS(L, P \cup \{p \leftarrow | p \in B\})$. Let c (resp., d^-) be the sum of all $\mu(B)$ such that (i) B is a total choice of C and (ii) $\alpha \wedge \neg\beta$ is true (resp., false) in $WFS(L,\ P \cup \{p \leftarrow | p \in B\})$. Let $b = 1 - b^-$ and $d = 1 - d^-$. Then, *the tight answer θ for Q to KB under the total well-founded semantics* (denoted $TWFS(KB)$) is defined as follows:

$$\theta = \begin{cases} \{r/1,\, s/0\} & \text{if } b=0 \text{ and } d=0; \\ \{r/0,\, s/0\} & \text{if } b=0 \text{ and } d\neq 0; \\ \{r/1,\, s/1\} & \text{if } b\neq 0 \text{ and } d=0; \\ \{r/\frac{a}{a+d},\, s/\frac{b}{b+c}\} & \text{otherwise.} \end{cases}$$

The following theorem shows that for probabilistic queries $Q = \exists(\ell)[r,s]$, where ℓ is a ground literal, the tight answers under the total well-founded semantics approximate the tight answers under the answer set semantics (if they exist). This is a nice semantic feature of the total well-founded semantics. It allows for an efficient approximation of tight answers to such queries under the answer set semantics by the bottom-up fixpoint iteration of the well-founded semantics of normal dl-programs.

Theorem 3. *Let $KB = (L, P, C, \mu)$ be a consistent probabilistic dl-program, and let $Q = \exists(\ell)[r,s]$ be a probabilistic query with ground literal ℓ. Let $\theta = \{r/l,\ s/u\}$ (resp., $\theta' = \{r/l',\ s/u'\}$) be the tight answer for Q to KB under the total well-founded semantics (resp., answer set semantics). Then, $[l', u'] \subseteq [l, u]$.*

The next theorem shows that the total well-founded semantics generalizes the well-founded semantics of [53], i.e., the tight answers under the former coincide with the tight answers under the latter, if the tight answers under the latter exist.

Theorem 4. *Let $KB = (L, P, C, \mu)$ be a probabilistic dl-program, and let $Q = \exists(\beta\,|\,\alpha)[r,s]$ be a probabilistic query with ground $\beta|\alpha$. Then, the tight answer for Q to KB under the total well-founded semantics coincides with the tight answer for Q to KB under the well-founded semantics of [53] (if it exists).*

7 Algorithms and Complexity

In this section, we provide an anytime algorithm for tight query processing in probabilistic dl-programs under the total well-founded semantics, and conclude with tractability and complexity results.

7.1 An Anytime Algorithm for Tight Query Processing

By Definition 1, computing the tight answer for a probabilistic query to a probabilistic dl-program $KB = (L, P, C, \mu)$ under $TWFS(KB)$ can be reduced to computing the well-founded models of all normal dl-programs $(L, P \cup \{p \leftarrow | \, p \in B\})$ such that B is a total choice of C. Here, the number of all total choices B is generally a non-neglectable source of complexity. We thus propose:

- to compute the tight answer only up to an error within a given threshold $\varepsilon \in [0,1]$,
- to process the B's along decreasing probabilities $\mu(B)$, and
- to eventually stop the computation after a given time interval.

Given a (not necessarily consistent) probabilistic dl-program $KB = (L, P, C, \mu)$, a probabilistic query $Q = \exists(\beta|\alpha)[r,s]$ with ground $\beta|\alpha$, and an error threshold $\varepsilon \in [0,1]$, algorithm tight_answer (see Fig. 3) computes some $\theta = \{r/l', \, s/u'\}$ such that $|l - l'| + |u - u'| \leqslant \varepsilon$, where $\{r/l, s/u\}$ is the tight answer for Q to KB under $TWFS(KB)$. More concretely, it computes the bounds l' and u' by first initializing the variables a, b, c, and d (which play the same role as in Definition 1). It then computes the well-founded semantics S of the normal dl-program $(L, P \cup \{p \leftarrow | \, p \in B_i\})$ for every total choice B_i of C, checks whether $\alpha \wedge \beta$ and $\alpha \wedge \neg\beta$ are true or false in S, and updates a, b, c, and d accordingly. If the possible error in the bounds falls below ε, then it stops and returns the bounds computed so far. Thus, in the special case where $\varepsilon = 0$, the algorithm computes in particular the tight answer for Q to KB under $TWFS(KB)$. The following theorem shows that algorithm tight_answer is sound.

Theorem 5. *Let KB be a probabilistic dl-program, let $Q = \exists(\beta|\alpha)[r,s]$ be a probabilistic query with ground $\beta|\alpha$, and let $\theta = \{r/l, s/u\}$ be the tight answer for Q to KB under TWFS(KB). Let $\theta' = \{r/l', s/u'\}$ be the output computed by tight_answer for the error threshold $\varepsilon \in [0,1]$. Then, $|l - l'| + |u - u'| \leqslant \varepsilon$.*

Algorithm tight_answer is actually an *anytime algorithm*, since we can always interrupt it, and return the bounds computed thus far. The following theorem shows that these bounds deviate from the tight bounds with an exactly measurable error (note that the possible error is also decreasing along the iterations of the while-loop). For this reason, algorithm tight_answer also iterates through the total choices B_i of C in a way such that the probabilities $\mu(B_i)$ are decreasing, so that the error in the computed bounds is very likely to be low already after few iteration steps.

Theorem 6. *Let KB be a probabilistic dl-program, let $Q = \exists(\beta|\alpha)[r,s]$ be a probabilistic query with ground $\beta|\alpha$, let $\varepsilon \in [0,1]$ be an error threshold, and let $\theta = \{r/l, s/u\}$ be the tight answer for Q to KB under TWFS(KB). Suppose we run tight_answer on KB, Q, and ε, and we interrupt it after line (9). Let the returned $\theta' = \{r/l', s/u'\}$ be as specified in lines (11) to (14). Then, if $v = 0$, then $\theta = \theta'$. Otherwise,*

$$|l - l'| + |u - u'| \leqslant \frac{v}{a+d} + \frac{v}{b+c}.$$

The algorithm is based on two finite fixpoint iterations for computing the well-founded semantics of normal dl-programs, which are in turn based on a finite fixpoint

Algorithm tight_answer

Input: probabilistic dl-program $KB = (L, P, C, \mu)$, probabilistic query $Q = \exists(\beta|\alpha)[r,s]$ with ground $\beta|\alpha$, and error threshold $\varepsilon \in [0,1]$.

Output: $\theta = \{r/l', s/u'\}$ such that $|l - l'| + |u - u'| \leqslant \varepsilon$, where $\{r/l, s/u\}$ is the tight answer for Q to KB under the total well-founded semantics.

Notation: $B_1, \ldots, B_k$ is a sequence of all total choices B of C with $\mu(B_1) \geqslant \cdots \geqslant \mu(B_k)$.

$a := 0;\ b := 1;\ c := 0;\ d := 1;\ v := 1;\ i := 1;$
while $i \leqslant k$ and $v > 0$ and $\frac{v}{a+d} + \frac{v}{b+c} > \varepsilon$ **do begin**
  $S := WFS(L, P \cup \{p \leftarrow \mid p \in B_i\});$
  if $\alpha \wedge \beta$ is true in S **then** $a := a + \mu(B_i)$
    else if $\alpha \wedge \beta$ is false in S **then** $b := b - \mu(B_i);$
  if $\alpha \wedge \neg\beta$ is true in S **then** $c := c + \mu(B_i)$
    else if $\alpha \wedge \neg\beta$ is false in S **then** $d := d - \mu(B_i);$
  $v := v - \mu(B_i);$
  $i := i + 1$
end;
if $b = 0$ and $d = 0$ **then return** $\theta = \{r/1, s/0\}$
  else if $b = 0$ and $d \neq 0$ **then return** $\theta = \{r/0, s/0\}$
    else if $b \neq 0$ and $d = 0$ **then return** $\theta = \{r/1, s/1\}$
      else return $\theta = \{r/\frac{a}{a+d}, s/\frac{b}{b+c}\}$.

Fig. 3 Algorithm tight_answer

iteration for computing the least model of positive dl-programs. More specifically, to compute the well-founded semantics of KB, i.e.,

$$WFS(KB) = lfp(\gamma^2_{KB}) \cup \neg(HB_\Phi - gfp(\gamma^2_{KB})),$$

we compute $lfp(\gamma^2_{KB})$ and $gfp(\gamma^2_{KB})$ as the limits of the two finite fixpoint iterations

$$U_0 = \emptyset, \text{ and } U_{i+1} = \gamma^2_{KB}(U_i), \text{ for } i \geqslant 0, \text{ and}$$
$$O_0 = HB_\Phi, \text{ and } O_{i+1} = \gamma^2_{KB}(O_i), \text{ for } i \geqslant 0,$$

respectively. Here, the operator γ_{KB}, which is defined by $\gamma_{KB}(I) = M_{KB^I}$ (with $KB^I = (L, P^I_L)$) for all $I \subseteq HB_\Phi$, is computed as the limit of the finite fixpoint iteration

$$S_0 = \emptyset, \text{ and } S_{i+1} = T_{KB^I}(S_i), \text{ for } i \geqslant 0,$$

since $\gamma_{KB}(I) = lfp(T_{KB^I})$ for all $I \subseteq HB_\Phi$, where T_{KB^I} is the immediate consequence operator for positive dl-programs, which is defined as follows for every $J \subseteq HB_\Phi$:

$$T_{KB^I}(J) = \{H(r) \mid r \in ground(P^I_L), J \models_L \ell \text{ for all } \ell \in B(r)\}.$$

All the above three fixpoint iterations are finite, since HB_Φ is finite.

7.2 Complexity

The following theorem shows that tight query processing in probabilistic dl-programs $KB = (L, P, C, \mu)$ in *DL-Lite* (i.e., L is in *DL-Lite*) under $TWFS(KB)$ can be done in polynomial time in the data complexity. This follows from Theorem 5 and the polynomial data complexity of (a) computing the well-founded semantics of a normal dl-program (see above) and of (b) conjunctive query processing in *DL-Lite*. Here, $|C|$ is bounded by a constant, since C and μ define the probabilistic information of P, which is fixed as a part of the program in P, while the ordinary facts in P (along with the concept and role membership axioms in L) are the variable input.

Theorem 7. *Given a probabilistic dl-program KB in DL-Lite and a probabilistic query* $Q = \exists(\ell)[r, s]$ *with ground literal* ℓ*, the tight answer* $\theta = \{r/l, s/u\}$ *for Q to KB under* $TWFS(KB)$ *can be computed in polynomial time in the data complexity.*

The next theorem shows that computing tight answers is EXP-complete in the combined complexity. The lower bound follows from the EXP-hardness of Datalog in the combined complexity, and the upper bound follows from Theorem 5.

Theorem 8. *Given a probabilistic dl-program KB in DL-Lite and a probabilistic query* $Q = \exists(\beta|\alpha)[r, s]$ *with ground* $\beta|\alpha$*, computing the tight answer* $\theta = \{r/l, s/u\}$ *for Q to KB under* $TWFS(KB)$ *is* EXP*-complete in the combined complexity.*

8 Probabilistic Data Integration

Integrating data from different sources is a crucial issue in the Semantic Web. In this section, we show how probabilistic dl-programs can be employed as a formalism for data integration in the Semantic Web. We first give some general definitions.

A *data integration system* (in its most general form, see [47]) $I = (G, S, M)$ consists of the following components:

- a *global* (or *mediated*) *schema* G, which represents the domain of interest,
- a *source schema* S, which represents the data sources of the system, and
- a *mapping* M, which relates the source schema and the global schema.

Here, G is purely virtual, while the data are stored in S. The mapping M can be specified in different ways, which is a crucial aspect in a data integration system. In particular, when every data structure in G is defined through a view over S, the mapping is said to be *GAV (global-as-view)*, while when every data structure in S is defined through a view over G the mapping is *LAV (local-as-view)*. A mixed approach, called *GLAV* [28, 12], associates views over G to views over S.

8.1 Modeling Data Integration Systems

In our framework, we assume that the global schema G, the source schema S, and the mapping M are each encoded by a probabilistic dl-program. More formally, we partition the vocabulary Φ into the sets Φ_G, Φ_S, and Φ_c such that:

- the symbols in Φ_G are of arity at least 1 and represent the global predicates,
- the symbols in Φ_S are of arity at least 1 and represent source predicates, and
- the symbols in Φ_c are constants.

Let $\mathbf{A}_G$ and $\mathbf{R}_G$ be disjoint denumerable sets of atomic concepts and abstract roles, respectively, for the global schema, and let $\mathbf{A}_S$ and $\mathbf{R}_S$ (disjoint from $\mathbf{A}_G$ and $\mathbf{R}_G$) be similar sets for the source schema. We also assume a denumerable set of individuals $\mathbf{I}$ that is disjoint from the set of all concepts and roles and a superset of Φ_c. A *probabilistic data integration system* $PI = (KB_G, KB_S, KB_M)$ consists of a probabilistic dl-program $KB_G = (L_G, P_G, C_G, \mu_G)$ for the global schema, a probabilistic dl-program $KB_S = (L_S, P_S, C_S, \mu_S)$ for the source schema, and a probabilistic dl-program $KB_M = (\emptyset, P_M, C_M, \mu_M)$ for the mapping:

- KB_G (resp., KB_S) is defined over the predicates, constants, concepts, roles, and individuals of the global (resp., source) schema, and it encodes ontological, rule-based, and probabilistic relationships in the global (resp., source) schema.
- KB_M is defined over the predicates, constants, concepts, roles, and individuals of the global and the source schema, and it encodes a probabilistic mapping between the predicates, concepts, and roles of the source and those of the global schema.

Our probabilistic dl-rules permit a specification of the mapping that can freely use global and source predicates together in rules, thus having a formalism that generalizes LAV and GAV in some way. Moreover, with a simple technicality, we are able to partly model GLAV systems. In GLAV data integration systems, the mapping is specified by means of rules of the form $\psi \leftarrow \varphi$, where ψ is a conjunction of atoms of G, and φ is a conjunction of atoms of S. We introduce an auxiliary atom α that contains all the variables of ψ; moreover, let $\psi = \beta_1 \wedge \ldots \wedge \beta_m$. We model the GLAV mapping rule with the following rules:

$$\begin{aligned} \beta_1 &\leftarrow \alpha; \\ &\vdots \\ \beta_m &\leftarrow \alpha; \\ \alpha &\leftarrow \varphi. \end{aligned}$$

What our framework does not allow is having rules that are *unsafe*, i.e., having existentially-quantified variables in their head.

Note also that correct and tight answers to probabilistic queries on the global schema are formally defined relative to the probabilistic dl-program $KB = (L, P, C, \mu)$, where $L = L_G \cup L_S$, $P = P_G \cup P_S \cup P_M$, $C = C_G \cup C_S \cup C_M$, and $\mu = \mu_G \cdot \mu_S \cdot \mu_M$. Informally, KB is the result of merging KB_G, KB_S, and KB_M. In a similar way, the probabilistic dl-program KB_S of the source schema S can be defined by merging the probabilistic dl-programs $KB_{S_1}, \ldots, KB_{S_1}$ of $n \geqslant 1$ source schemas $S_1, \ldots, S_n$.

The fact that the mapping is probabilistic allows for a high flexibility in the treatment of the uncertainty that is present when pieces of data come from heterogeneous sources whose content may be inconsistent and/or redundant relative to the global schema G, which in general incorporates constraints. Some different types of probabilistic mappings that can be modeled in our framework are summarized below.

8.2 Types of Probabilistic Mappings

In addition to expressing probabilistic knowledge about the global schema and about the source schema, the probabilities in probabilistic dl-programs can especially be used for specifying the probabilistic mapping in the data integration process. We distinguish three different types of probabilistic mappings, depending on whether the probabilities are used as *trust*, *error*, or *mapping probabilities*.

The simplest way of probabilistically integrating several data sources is to weight each data source with a *trust probability* (which all sum up to 1). This is especially useful when several redundant data sources are to be integrated. In such a case, pieces of data from different data sources may easily be inconsistent with each other.

Example 7. Suppose that we want to obtain a weather forecast for a certain place by integrating the potentially different weather forecasts of several weather forecast institutes. For ease of presentation, suppose that we only have three weather forecast institutes A, B, and C. In general, one trusts certain weather forecast institutes more than others. In our case, we suppose that our trust in the institutes A, B, and C is expressed by the trust probabilities 0.6, 0.3, and 0.1, respectively. That is, we trust most in A, medium in B, and less in C. In general, the different institutes do not use the same data structure to represent their weather forecast data. For example, institute A may use a single relation

$$\mathit{forecast}(\mathit{place}, \mathit{date}, \mathit{weather}, \mathit{temperature}, \mathit{wind})$$

to store all the data, while B may have one relation

$$\mathit{forecast\_place}(\mathit{date}, \mathit{weather}, \mathit{temperature}, \mathit{wind})$$

for each place, and C may use several different relations

$$\begin{array}{l}\mathit{forecast\_weather}(\mathit{place}, \mathit{date}, \mathit{weather}),\\ \mathit{forecast\_temperature}(\mathit{place}, \mathit{date}, \mathit{temperature}),\\ \mathit{forecast\_wind}(\mathit{place}, \mathit{date}, \mathit{wind}).\end{array}$$

Suppose the global schema G has the relation

$$\mathit{forecast\_rome\_global}(\mathit{date}, \mathit{weather}, \mathit{temperature}, \mathit{wind}),$$

which may for instance be posted on the Web by the tourist information of the city of Rome. The probabilistic mapping of the source schemas of A, B, and C to the global schema G can then be specified by the following $KB_M = (\emptyset, P_M, C_M, \mu_M)$:

$$\begin{aligned}P_M = \{&\mathit{forecast\_rome\_global}(D,W,T,M) \leftarrow \mathit{forecast}(\mathit{rome},D,W,T,M),\ \mathit{inst}_A;\\ &\mathit{forecast\_rome\_global}(D,W,T,M) \leftarrow \mathit{forecast\_rome}(D,W,T,M),\ \mathit{inst}_B;\\ &\mathit{forecast\_rome\_global}(D,W,T,M) \leftarrow \mathit{forecast\_weather}(\mathit{rome},D,W),\\ &\quad \mathit{forecast\_temperature}(\mathit{rome},D,T), \mathit{forecast\_wind}(\mathit{rome},D,M),\ \mathit{inst}_C\};\\ C_M = \{&\{\mathit{inst}_A, \mathit{inst}_B, \mathit{inst}_C\}\};\\ \mu_M :\ &\mathit{inst}_A, \mathit{inst}_B, \mathit{inst}_C \mapsto 0.6, 0.3, 0.1\,.\end{aligned}$$

The mapping assertions state that the first, second, and third rule above hold with the probabilities 0.6, 0.3, and 0.1, respectively. This is motivated by the fact that three institutes may generally provide conflicting weather forecasts, and our trust in A, B, and C are given by the trust probabilities 0.6, 0.3, and 0.1, respectively.

A more complex way of probabilistically integrating several data sources is to associate each data source (or each derivation) with an *error probability*.

Example 8. Suppose that we want to integrate the data provided by the different sensors in a sensor network. For example, assume a sensor network measuring the concentration of ozone in several different positions of a certain town, which may for instance be the basis for the common hall to reduce or forbid individual traffic. Suppose that each sensor $i \in \{1, \ldots, n\}$ with $n \geqslant 1$ is associated with its position through $sensor(i, position)$ and provides its measurement data in a single relation $reading_i(date, time, type, result)$. Each such reading may be erroneous with the probability e_i. That is, any tuple returned (resp., not returned) by a sensor $i \in \{1, \ldots, n\}$ may not hold (resp., may hold) with probability e_i. Let the global schema contain a single relation $reading(position, date, time, type, result)$. Then, the probabilistic mapping of the source schemas of the sensors $i \in \{1, \ldots, n\}$ to the global schema G can be specified by the following probabilistic dl-program $KB_M = (\emptyset, P_M, C_M, \mu_M)$:

$$\begin{aligned}
P_M = \; & \{aux_i(P, D, T, K, R) \leftarrow reading_i(D, T, K, R),\, sensor(i, P) \mid i \in \{1, \ldots, n\}\} \,\cup \\
& \{reading(P, D, T, K, R) \leftarrow aux_i(P, D, T, K, R),\, not_error_i \mid i \in \{1, \ldots, n\}\} \,\cup \\
& \{reading(P, D, T, K, R) \leftarrow not\, aux_i(P, D, T, K, R),\, error_i \mid i \in \{1, \ldots, n\}\};
\end{aligned}$$

$$C_M = \{\{error_i, not\_error_i\} \mid i \in \{1, \ldots, n\}\};$$

$$\mu_M : \; error_1, not\_error_1, \ldots, error_n, not\_error_n \;\mapsto\; e_1,\, 1-e_1, \ldots,\, e_n,\, 1-e_n\,.$$

Note that if there are two sensors j and k for the same position, and they both return the same tuple as a reading, then this reading is correct with the probability $1 - e_j e_k$ (since it may be erroneous with the probability $e_j e_k$). Note also that this modeling assumes that the errors of the sensors are independent from each other, which can be achieved by eventually unifying atomic choices. For example, if the sensor j depends on the sensor k, then j is erroneous when k is erroneous, and thus the atomic choices $\{error_j, not\_error_j\}$ and $\{error_k, not\_error_k\}$ are merged into the new atomic choice $\{error_j error_k,\, not\_error_j error_k,\, not\_error_j not\_error_k\}$.

When integrating several data sources, it may be the case that the relationships between the source schema and the global schema are *purely probabilistic*.

Example 9. Suppose that we want to integrate the schemas of two libraries, and that the global schema contains the predicate symbol *logic_programming*, while the source schemas contain only the concepts *rule-based_systems* and *deductive_databases* in their ontologies. These three concepts are overlapping to some extent, but they do not exactly coincide. For example, a randomly chosen book from *rule-based_systems* (resp., *deductive_databases*) may belong to *logic_programming* with the probability 0.7 (resp., 0.8). The probabilistic mapping from the source schemas to the global schema can then be expressed by the following $KB_M = (\emptyset, P_M, C_M, \mu_M)$:

$$
\begin{aligned}
P_M &= \{\mathit{logic\_programming}(X) \leftarrow DL[\mathit{rule\text{-}based\_systems}(X)],\, \mathit{choice}_1\,;\\
&\qquad \mathit{logic\_programming}(X) \leftarrow DL[\mathit{deductive\_databases}(X)],\, \mathit{choice}_2\}\,;\\
C_M &= \{\{\mathit{choice}_1, \mathit{not\_choice}_1\}, \{\mathit{choice}_2, \mathit{not\_choice}_2\}\}\,;\\
\mu_M &: \ \mathit{choice}_1, \mathit{not\_choice}_1, \mathit{choice}_2, \mathit{not\_choice}_2 \ \mapsto\ 0.7, 0.3, 0.8, 0.2\,.
\end{aligned}
$$

8.3 Deterministic Mappings on Probabilistic Data

Finally, we briefly describe an approach to use probabilistic dl-programs to model probabilistic data, such as those in [17].

Example 10. Suppose that the weather in Oxford can be sunny, cloudy, or rainy with probabilities 0.2, 0.45, and 0.35, respectively, and similar probabilities are assigned for other cities. This setting is analogous to the "classical" one of probabilistic data, where there is a probability distribution over ground facts. In such a case, the choice space is $C = \{\{\mathit{weather}(\mathit{oxford},\ \mathit{sunny}),\ \mathit{weather}(\mathit{oxford},\ \mathit{cloudy}),\ \mathit{weather}(\mathit{oxford},\ \mathit{rainy})\}, \ldots\}$, and the probability is μ: $\mathit{weather}(\mathit{oxford},\ \mathit{sunny})$, $\mathit{weather}(\mathit{oxford},\ \mathit{cloudy})$, $\mathit{weather}(\mathit{oxford},\ \mathit{rainy}) \mapsto 0.2, 0.45, 0.35$. A mapping rule such as

$$\mathit{candidate\_destination}(L) \leftarrow \mathit{weather}(L, \mathit{sunny})$$

can now express the fact that a destination is a candidate for a day-trip if it has sunny weather. While the mapping is purely deterministic, the probability distributions on the sets of atomic choices of the choice space enforce, by virtue of the mapping, a probability distribution on the ground facts of the global schema. Our framework is able to capture this situation, allowing for query answering over uncertain data.

9 Related Work

In this section, we discuss more closely related work on (a) the combination of logic programs, DLs, and probabilistic uncertainty, (b) the combination of logic programs and probabilistic uncertainty, (c) the combination of logic programs and DLs, and (d) the combination of DLs (or ontology languages) and probabilistic uncertainty. Note that more detailed overviews on uncertainty reasoning for the Semantic Web, which covers (a), (c), and (d), are given in [59, 56].

9.1 Probabilistic Description Logic Programs

To our knowledge, the work of [53] was the first one combining (normal) dl-programs (under the loose integration) with probabilistic uncertainty. Instead of being based on the loosely integrated normal dl-programs $KB = (L, P)$ of [25, 26], probabilistic dl-programs can also be developed as a generalization of the *tightly integrated* ones in [50] (see [14, 57]). Rather than having dl-queries to L in rule bodies in P (which also allow for passing facts as dl-query arguments from P to L) and assuming that Φ and $\mathbf{A}$ (resp., $\mathbf{R}$) have no unary (resp., binary) predicate symbols in common (and so that dl-queries are the only interface between L and P), the tightly integrated normal dl-programs of [50] have no dl-queries, but Φ and $\mathbf{A}$ (resp., $\mathbf{R}$) may very well have unary (resp., binary) predicate symbols in common, and so the integration between L and P is of a much tighter nature. Nearly all the results of this paper carry over to such tightly integrated probabilistic dl-programs. As an important feature for the Semantic Web, they also allow for expressing in P probabilistic relations between the concepts and roles in L, since we can freely use concepts and roles from L as unary resp. binary predicate symbols in P.

The (loosely coupled) probabilistic fuzzy dl-programs in [58] combine fuzzy DLs, fuzzy logic programs (with stratified default-negation), and probabilistic uncertainty in a uniform framework for the Semantic Web. Intuitively, they allow for defining several rankings on ground atoms using fuzzy vagueness, and then for merging these rankings using probabilistic uncertainty (by associating with each ranking a probabilistic weight and building the weighted sum of all rankings). Less closely related, since they deal with fuzzy vagueness alone, rather than probabilistic ambiguity and imprecision, are the loosely and tightly coupled fuzzy dl-programs that have been introduced in [52] and [60], respectively, and extended by a top-k retrieval technique in [61]. Related works by Straccia combine (positive) dl-programs with lattice-based uncertainty [81] and with fuzzy vagueness [78].

9.2 Probabilistic Logic Programs

Ng and Subrahmanian [63, 62] proposed the first approach to probabilistic logic programs that addresses the problem of combining logic programs [49] with probability theory by adopting semantics in the style of Nilsson [65] and Halpern [35] for probabilistic logic. All semantics proposed for quantitative logic programming prior to this work had been non-probabilistic, of which [84] and [75] are examples; on the other hand, this approach aims at developing a probabilistic model theory and fixpoint theory. The general form of rules in their formalism is:

$$F_0 : \mu_0 \leftarrow F_1 : \mu_1 \wedge \ldots \wedge F_n : \mu_n \,,$$

where the F_i's are *basic formulas* (conjunctions or disjunctions of atoms) and the μ_i's are probabilistic annotations in the form of intervals that may contain expressions with variables. In [62], heads of rules are restricted to annotated atoms, where negation is still supported via $[0,0]$ annotations, but conditional probabilities are not expressible. The formalism is a general framework for expressing probabilistic information, and the authors study its semantics and relationship with probability theory, model theory, fixpoint theory, and proof theory. They also develop a procedure for answering queries about probabilities of events, which is different from query processing in classical logic programming, since most general unifiers are not always unique, and thus *maximally general* unifiers must be computed.

Other approaches to probabilistic logic programs have especially been proposed by Ngo and Haddawy [64], Lukasiewicz [54], Lukasiewicz and Kern-Isberner [41], Lakshmanan and Shiri [46], Dekhtyar and Subrahmanian [21], and Damasio et al. [19]. Ngo and Haddawy [64] present a model theory, fixpoint theory, and proof procedure for conditional probabilistic logic programming. Lukasiewicz [54] presents a conditional semantics for probabilistic logic programs where each rule is interpreted as specifying the conditional probability of the rule head, given the body. In closely related work, Lukasiewicz and Kern-Isberner [41] combine probabilistic logic programs (also adopting an explicit treatment of conditional probabilities) with maximum entropy, in relation to Nilsson's proposal for probabilistic logic [65]. Lakshmanan and Shiri [46] developed a semantics for logic programs in which different general axiomatic methods are given to compute probabilities of conjunctions and disjunctions, and these are used to define a semantics for probabilistic logic programs. In [21], Dekhtyar and Subrahmanian consider different conjunction and disjunction strategies, originally introduced by Lakshmanan et al. [45], and allow an explicit syntax in probabilistic logic programs so that users are able to express their knowledge of a dependency. Damasio et al. [19] present a well-founded semantics for annotated logic programs and show how to compute it.

Although there is a rich body of work on probabilistic logic programs, most such works to date have only addressed the problem of checking whether a given formula of the form $F\colon [\ell, u]$ is entailed by a probabilistic logic program [63, 62] or is true in a specific model (e.g., the well-founded model [19]). This usually boils down to finding out if all interpretations that satisfy the probabilistic logic program assign a probability between ℓ and u to F. An interesting extension of the concept of probabilistic entailment is proposed in [88], where the authors propose to go one step further and check to what *degree of satisfaction* the query is entailed by the program, similar to the novel approach of Bröcheler et al. [10], who propose to answer entailment queries with histograms indicating how the density of solutions are distributed in the probabilistic interval. In contrast, other works have recently focused on finding *most probable worlds* [43, 77], and answering *abductive queries* [76].

9.3 Description Logic Programs

Related work on the combination of logic programs and DLs can be divided into the following categories: (a) hybrid approaches using DLs as input to logic programs, (b) approaches reducing DLs to logic programs, (c) combinations of DLs with default and defeasible logic, and (d) approaches to rule-based well-founded reasoning in the Semantic Web. Below we give some representatives for these categories; further works and details are given in [25, 26, 50].

The works by Donini et al. [23], Levy and Rousset [48], and Rosati [73, 74] are representatives of hybrid approaches using DLs as input. Donini et al. [23] introduce a combination of (disjunction-, negation-, and function-free) Datalog with the DL $\mathcal{ALC}$. An integrated knowledge base consists of a structural component in $\mathcal{ALC}$ and a relational component in Datalog, where the integration of both components lies in using concepts from the structural component as constraints in rule bodies of the relational component. The closely related work by Levy and Rousset [48] presents a combination of Horn rules with the DL $\mathcal{ALC}$. In contrast to Donini et al. [23], Levy and Rousset also allow for roles as constraints in rule bodies, and do not require the safety condition that variables in constraints in the body of a rule r must also appear in ordinary atoms in the body of r. Finally, Rosati [73] presents a combination of disjunctive Datalog (with classical and default negation, but without function symbols) with $\mathcal{ALC}$, which is based on a generalized answer set semantics. Some approaches reducing DL reasoning to logic programming are the works by Van Belleghem et al. [6], Alsaç and Baral [1], Swift [82], Grosof et al. [34], and Hufstadt et al. [39]. Early work on dealing with default information in DLs is the approach due to Baader and Hollunder [4], where Reiter's default logic is adapted to terminological knowledge bases. Antoniou [2] combines defeasible reasoning with DLs for the Semantic Web. In [3], Antoniou and Wagner summarize defeasible and strict reasoning in a single rule formalism. An important approach to rule-based reasoning under the well-founded semantics for the Semantic Web is due to Damasio [18]. He aims at Prolog tools for implementing different semantics for RuleML [9]. So far, an XML parser library as well as a RuleML compiler have been developed, with routines to convert RuleML rule bases to Prolog and vice versa. The compiler supports paraconsistent well-founded semantics with explicit negation; it is planned to be extended to use XSB [71].

9.4 Probabilistic Description Logics and Ontology Languages

Probabilistic generalizations of the expressive DLs $\mathcal{SHOQ}(\mathbf{D})$, $\mathcal{SHIF}(\mathbf{D})$, and $\mathcal{SHOIN}(\mathbf{D})$ behind DAML+OIL, OWL Lite, and OWL DL, respectively, have been proposed by Giugno and Lukasiewicz [31] and Lukasiewicz [51]. They are based on lexicographic probabilistic reasoning. A companion paper [20] combines *DL-Lite* with Bayesian networks. In earlier works, Heinsohn [36] and Jaeger [40] present probabilistic extensions to the DL $\mathcal{ALC}$, which are essentially based on probabilistic reasoning in probabilistic logics. Koller et al. [44] present a probabilistic generalization of the CLASSIC DL, which uses Bayesian networks as underlying probabilistic reasoning formalism. Recently, an extension to the Datalog$^{\pm}$ family of ontology languages [13] has been developed in [32, 33] to add the capability of representing probabilistic uncertainty by means of an integration between Datalog$^{\pm}$ ontologies and Markov logic networks [72], focusing on scalability towards applications in data extraction and reasoning in the Web. Note that fuzzy DLs, such as the ones by Straccia [79, 80] are less closely related to probabilistic DLs, since they deal with fuzzy vagueness, rather than probabilistic ambiguity and imprecision.

Especially the works by Costa [16], Pool and Aikin [69], and Ding and Peng [22] present probabilistic generalizations of the Web ontology language OWL. In particular, Costa's work [16] is semantically based on multi-entity Bayesian networks, while [22] has a semantics in standard Bayesian networks. In closely related work, Fukushige [29] proposes a basic framework for representing probabilistic relationships in RDF. Finally, Nottelmann and Fuhr [66] present pDAML+OIL, which is a probabilistic generalization of the Web ontology language DAML+OIL, along with a mapping to stratified probabilistic Datalog.

10 Conclusion

We have proposed tractable probabilistic dl-programs for the Semantic Web, which combine tractable DLs, normal programs under the answer set and the well-founded semantics, and probabilities. We have given novel reductions of tight query processing and deciding consistency in probabilistic dl-programs under the answer set semantics to the answer set semantics of the underlying normal dl-programs. Based on them, we have then introduced the total well-founded semantics for probabilistic dl-programs. Contrary to the previous answer set and well-founded semantics, it is defined for all probabilistic dl-programs and queries. Furthermore, tight (resp., tight literal) query processing under the total well-founded semantics coincides with (resp., approximates) tight (resp., tight literal) query processing under the previous well-founded (resp., answer set) semantics in all cases where the latter is defined. We have then presented an anytime algorithm for tight query processing in probabilistic dl-programs under the total well-founded semantics. Note that the novel reductions, the total well-founded semantics, and the anytime algorithm are not limited to *DL-Lite* as underlying DL; they hold for all probabilistic dl-programs on top of DLs with decidable conjunctive query processing. We have also shown that tight query processing in probabilistic dl-programs under the total well-founded semantics is possible in polynomial time in the data complexity and is complete for EXP in the combined complexity. Finally, we have described an application of probabilistic dl-programs in probabilistic data integration for the Semantic Web.

An interesting topic for future research is to investigate whether one can also develop an efficient top-k query technique for the presented probabilistic dl-programs: Rather than computing the tight probability interval for a given ground literal, such a technique returns k most probable ground instances of a given non-ground formula.

Acknowledgements. This work was partially supported by the European Research Council under the European Union's 7th Framework Programme (FP7/2007-2013)/ERC grant 246858 — DIADEM, the EPSRC grant EP/J008346/1 "PrOQAW: Probabilistic Ontological Query Answering on the Web", a Yahoo! Research Fellowship, and a Google Research Award. This article is a significantly extended and revised version of a paper that appeared in *Proc. SUM-2007* [55].

References

1. Alsaç, G., Baral, C.: Reasoning in description logics using declarative logic programming. Technical report, Arizona State University (2001)
2. Antoniou, G.: Nonmonotonic Rule Systems on Top of Ontology Layers. In: Horrocks, I., Hendler, J. (eds.) ISWC 2002. LNCS, vol. 2342, pp. 394–398. Springer, Heidelberg (2002)

3. Antoniou, G., Wagner, G.: Rules and Defeasible Reasoning on the Semantic Web. In: Schröder, M., Wagner, G. (eds.) RuleML 2003. LNCS, vol. 2876, pp. 111–120. Springer, Heidelberg (2003)
4. Baader, F., Hollunder, B.: Embedding defaults into terminological knowledge representation formalisms. J. Autom. Reasoning 14(1), 149–180 (1995)
5. Baral, C., Gelfond, M., Rushton, J.N.: Probabilistic Reasoning With Answer Sets. In: Lifschitz, V., Niemelä, I. (eds.) LPNMR 2004. LNCS (LNAI), vol. 2923, pp. 21–33. Springer, Heidelberg (2003)
6. Belleghem, K.V., Denecker, M., Schreye, D.D.: A strong correspondence between description logics and open logic programming. In: Proc. ICLP 1997, pp. 346–360. MIT Press (1997)
7. Berners-Lee, T.: Spinning the Semantic Web: Bringing the World Wide Web to Its Full Potential. MIT Press (2003)
8. Berners-Lee, T., Fischetti, M.: Weaving the Web. Harper, San Francisco (1999)
9. Boley, H., Tabet, S., Wagner, G.: Design rationale of RuleML: A markup language for Semantic Web rules. In: Proc. SWWS 2001, pp. 381–401 (2001)
10. Broecheler, M., Simari, G.I., Subrahmanian, V.S.: Using Histograms to Better Answer Queries to Probabilistic Logic Programs. In: Hill, P.M., Warren, D.S. (eds.) ICLP 2009. LNCS, vol. 5649, pp. 40–54. Springer, Heidelberg (2009)
11. Calvanese, D., De Giacomo, G., Lembo, D., Lenzerini, M., Rosati, R.: Tractable reasoning and efficient query answering in description logics: The *DL-Lite* family. J. Autom. Reasoning 39(3), 385–429 (2007)
12. Calì, A.: Reasoning in Data Integration Systems: Why LAV and GAV Are Siblings. In: Zhong, N., Raś, Z.W., Tsumoto, S., Suzuki, E. (eds.) ISMIS 2003. LNCS (LNAI), vol. 2871, pp. 562–571. Springer, Heidelberg (2003)
13. Calì, A., Gottlob, G., Lukasiewicz, T.: A general Datalog-based framework for tractable query answering over ontologies. J. Web Sem (2012) (in press)
14. Calì, A., Lukasiewicz, T., Predoiu, L., Stuckenschmidt, H.: Tightly coupled probabilistic description logic programs for the Semantic Web. J. Data Sem. 12, 95–130 (2009)
15. da Costa, P.C.G., Laskey, K.B.: PR-OWL: A framework for probabilistic ontologies. In: Proc. FOIS 2006, pp. 237–249. IOS Press (2006)
16. da Costa, P.C.G.: Bayesian Semantics for the Semantic Web. PhD thesis, Fairfax, VA, USA (2005)
17. Dalvi, N., Suciu, D.: Management of probabilistic data: foundations and challenges. In: Proc. PODS 2007, pp. 1–12. ACM Press (2007)
18. Damasio, C. V.: The W^4 project (2002), `http://centria.di.fct.unl.pt/~cd/projectos/w4/index.htm`
19. Viegas Damásio, C., Moniz Pereira, L., Swift, T.: Coherent Well-founded Annotated Logic Programs. In: Gelfond, M., Leone, N., Pfeifer, G. (eds.) LPNMR 1999. LNCS (LNAI), vol. 1730, pp. 262–276. Springer, Heidelberg (1999)
20. d'Amato, C., Fanizzi, N., Lukasiewicz, T.: Tractable Reasoning with Bayesian Description Logics. In: Greco, S., Lukasiewicz, T. (eds.) SUM 2008. LNCS (LNAI), vol. 5291, pp. 146–159. Springer, Heidelberg (2008)
21. Dekhtyar, A., Subrahmanian, V.S.: Hybrid probabilistic programs. In: Proc. ICLP 1997, pp. 391–405. MIT Press (1997)
22. Ding, Z., Peng, Y.: A probabilistic extension to ontology language OWL. In: Proc. HICSS 2004, IEEE Computer Society (2004)
23. Donini, F.M., Lenzerini, M., Nardi, D., Schaerf, A.: $\mathcal{AL}$-log: Integrating Datalog and description logics. J. Intell. Inf. Syst. 10(3), 227–252 (1998)

24. Eiter, T., Ianni, G., Schindlauer, R., Tompits, H.: A uniform integration of higher-order reasoning and external evaluations in answer-set programming. In: Proc. IJCAI 2005, pp. 90–96. Morgan Kaufmann (2005)
25. Eiter, T., Ianni, G., Lukasiewicz, T., Schindlauer, R., Tompits, H.: Combining answer set programming with description logics for the Semantic Web. Artif. Intell. 172(12/13), 1495–1539 (2008)
26. Eiter, T., Ianni, G., Lukasiewicz, T., Schindlauer, R.: Well-founded semantics for description logic programs in the Semantic Web. ACM Trans. Comput. Log. 12(2), 11 (2011)
27. Finzi, A., Lukasiewicz, T.: Structure-based causes and explanations in the independent choice logic. In: Proc. UAI 2003, pp. 225–232. Morgan Kaufmann (2003)
28. Friedman, M., Levy, A.Y., Millstein, T.D.: Navigational plans for data integration. In: Proc. AAAI 1999, pp. 67–73. AAAI Press (1999)
29. Fukushige, Y.: Representing probabilistic knowledge in the Semantic Web. In: Proceedings of the W3C Workshop on Semantic Web for Life Sciences (2004)
30. Gelfond, M., Lifschitz, V.: Classical negation in logic programs and disjunctive databases. New Generat. Comput. 9(3/4), 365–386 (1991)
31. Giugno, R., Lukasiewicz, T.: P-$\mathcal{SHOQ}(\mathbf{D})$: A Probabilistic Extension of $\mathcal{SHOQ}(\mathbf{D})$ for Probabilistic Ontologies in the Semantic Web. In: Flesca, S., Greco, S., Leone, N., Ianni, G. (eds.) JELIA 2002. LNCS (LNAI), vol. 2424, pp. 86–97. Springer, Heidelberg (2002)
32. Gottlob, G., Lukasiewicz, T., Simari, G.I.: Answering Threshold Queries in Probabilistic Datalog+/– Ontologies. In: Benferhat, S., Grant, J. (eds.) SUM 2011. LNCS, vol. 6929, pp. 401–414. Springer, Heidelberg (2011)
33. Gottlob, G., Lukasiewicz, T., Simari, G.I.: Conjunctive query answering in probabilistic datalog+/– ontologies. In: Rudolph, S., Gutierrez, C. (eds.) RR 2011. LNCS, vol. 6902, pp. 77–92. Springer, Heidelberg (2011)
34. Grosof, B.N., Horrocks, I., Volz, R., Decker, S.: Description logic programs: Combining logic programs with description logic. In: Proc. WWW 2003, pp. 48–57. ACM Press (2003)
35. Halpern, J.Y.: An analysis of first-order logics of probability. Artif. Intell. 46(3), 311–350 (1990)
36. Heinsohn, J.: Probabilistic description logics. In: Proc. UAI 1994, pp. 311–318. Morgan Kaufmann (1994)
37. Horrocks, I., Patel-Schneider, P.F.: Reducing OWL entailment to description logic satisfiability. J. Web Sem. 1(4), 345–357 (2004)
38. Horrocks, I., Patel-Schneider, P.F., van Harmelen, F.: From $\mathcal{SHIQ}$ and RDF to OWL: The making of a Web ontology language. J. Web Sem. 1(1), 7–26 (2003)
39. Hufstadt, U., Motik, B., Sattler, U.: Reasoning for description logics around $\mathcal{SHIQ}$ in a resolution framework. Technical report, FZI Karlsruhe (2004)
40. Jaeger, M.: Probabilistic reasoning in terminological logics. In: Proc. KR 1994, pp. 305–316. Morgan Kaufmann (1994)
41. Kern-Isberner, G., Lukasiewicz, T.: Combining probabilistic logic programming with the power of maximum entropy. Artif. Intell. 157(1/2), 139–202 (2004)
42. Kersting, K., De Raedt, L.: Bayesian logic programs. Technical report (2001)
43. Khuller, S., Martinez, M.V., Nau, D.S., Simari, G.I., Sliva, A., Subrahmanian, V.S.: Computing most probable worlds of action probabilistic logic programs: Scalable estimation for $10^{30,000}$ worlds. Ann. Math. Artif. Intell. 51(2-4), 295–331 (2007)
44. Koller, D., Levy, A.Y., Pfeffer, A.: P-CLASSIC: A tractable probabilistic description logic. In: Proc. AAAI 1997, pp. 390–397. AAAI Press, MIT Press (1997)

45. Lakshmanan, L.V.S., Leone, N., Ross, R., Subrahmanian, V.S.: ProbView: A flexible probabilistic database system. ACM Trans. Database Syst. 22(3), 419–469 (1997)
46. Lakshmanan, L.V.S., Shiri, N.: A parametric approach to deductive databases with uncertainty. IEEE Trans. Knowl. Data Eng. 13(4), 554–570 (2001)
47. Lenzerini, M.: Data integration: A theoretical perspective. In: Proc. PODS 2002, pp. 233–246. ACM Press (2002)
48. Levy, A.Y., Rousset, M.-C.: Combining Horn rules and description logics in CARIN. Artif. Intell. 104(1/2), 165–209 (1998)
49. Lloyd, J.W.: Foundations of Logic Programming, 2nd edn. Springer (1987)
50. Lukasiewicz, T.: A novel combination of answer set programming with description logics for the Semantic Web. IEEE Trans. Knowl. Data Eng. 22(11), 1577–1592 (2010)
51. Lukasiewicz, T.: Expressive probabilistic description logics. Artif. Intell. 172(6/7), 852–883 (2008)
52. Lukasiewicz, T.: Fuzzy description logic programs under the answer set semantics for the Semantic Web. Fundam. Inform. 82(3), 289–310 (2008)
53. Lukasiewicz, T.: Probabilistic description logic programs. Int. J. Approx. Reason. 45(2), 288–307 (2007)
54. Lukasiewicz, T.: Probabilistic logic programming with conditional constraints. ACM Trans. Comput. Log. 2(3), 289–339 (2001)
55. Lukasiewicz, T.: Tractable Probabilistic Description Logic Programs. In: Prade, H., Subrahmanian, V.S. (eds.) SUM 2007. LNCS (LNAI), vol. 4772, pp. 143–156. Springer, Heidelberg (2007)
56. Lukasiewicz, T.: Uncertainty Reasoning for the Semantic Web. In: Polleres, A., Swift, T. (eds.) RR 2009. LNCS, vol. 5837, pp. 26–39. Springer, Heidelberg (2009)
57. Lukasiewicz, T., Predoiu, L., Stuckenschmidt, H.: Tightly integrated probabilistic description logic programs for representing ontology mappings. Ann. Math. Artif. Intell. (2011)
58. Lukasiewicz, T., Straccia, U.: Description logic programs under probabilistic uncertainty and fuzzy vagueness. Int. J. Approx. Reasoning 50(6), 837–853 (2009)
59. Lukasiewicz, T., Straccia, U.: Managing uncertainty and vagueness in description logics for the Semantic Web. J. Web Sem. 6(4), 291–308 (2008)
60. Lukasiewicz, T., Straccia, U.: Tightly coupled fuzzy description logic programs under the answer set semantics for the Semantic Web. Int. J. Semantic Web Inf. Syst. 4(3), 68–89 (2008)
61. Lukasiewicz, T., Straccia, U.: Top-k Retrieval in Description Logic Programs Under Vagueness for the Semantic Web. In: Prade, H., Subrahmanian, V.S. (eds.) SUM 2007. LNCS (LNAI), vol. 4772, pp. 16–30. Springer, Heidelberg (2007)
62. Ng, R.T., Subrahmanian, V.S.: Probabilistic logic programming. Inform. Comput. 101(2), 150–201 (1992)
63. Ng, R.T., Subrahmanian, V.S.: A semantical framework for supporting subjective and conditional probabilities in deductive databases. J. Autom. Reasoning 10(2), 191–235 (1993)
64. Ngo, L., Haddawy, P.: Probabilistic logic programming and Bayesian networks. In: Kanchanasut, K., Levy, J.-J. (eds.) ACSC 1995. LNCS, vol. 1023, pp. 286–300. Springer, Heidelberg (1995)
65. Nilsson, N.J.: Probabilistic logic. Artif. Intell. 28(1), 71–87 (1986)
66. Nottelmann, H., Fuhr, N.: pDAML+OIL: A probabilistic extension to DAML+OIL based on probabilistic Datalog. In: Proc. IPMU 2004 (2004)
67. Pan, R., Ding, Z., Yu, Y., Peng, Y.: A Bayesian Network Approach to Ontology Mapping. In: Gil, Y., Motta, E., Benjamins, V.R., Musen, M.A. (eds.) ISWC 2005. LNCS, vol. 3729, pp. 563–577. Springer, Heidelberg (2005)

68. Poggi, A., Lembo, D., Calvanese, D., De Giacomo, G., Lenzerini, M., Rosati, R.: Linking data to ontologies. J. Data Semantics 10, 133–173 (2008)
69. Pool, M., Aikin, J.: KEEPER and Protégé: An elicitation environment for Bayesian inference tools. In: Proceedings of the Workshop on Protégé and Reasoning (2004)
70. Poole, D.: The independent choice logic for modelling multiple agents under uncertainty. Artif. Intell. 94(1/2), 7–56 (1997)
71. Rao, P., Sagonas, K.F., Swift, T., Warren, D.S., Freire, J.: XSB: A system for efficiently computing WFS. In: Fuhrbach, U., Dix, J., Nerode, A. (eds.) LPNMR 1997. LNCS, vol. 1265, pp. 431–441. Springer, Heidelberg (1997)
72. Richardson, M., Domingos, P.: Markov logic networks. Mach. Learn. 62(1/2), 107–136 (2006)
73. Rosati, R.: Towards expressive KR systems integrating Datalog and description logics: Preliminary report. In: Proc. DL 1999, pp. 160–164 (1999)
74. Rosati, R.: On the decidability and complexity of integrating ontologies and rules. J. Web Sem. 3(1), 61–73 (2005)
75. Shapiro, E.Y.: Logic programs with uncertainties: A tool for implementing rule-based systems. In: Proc. IJCAI 1983, pp. 529–532. William Kaufmann (1983)
76. Simari, G.I., Dickerson, J.P., Subrahmanian, V.S.: Cost-Based Query Answering in Action Probabilistic Logic Programs. In: Deshpande, A., Hunter, A. (eds.) SUM 2010. LNCS, vol. 6379, pp. 319–332. Springer, Heidelberg (2010)
77. Simari, G.I., Martinez, M.V., Sliva, A., Subrahmanian, V.S.: Focused most probable world computations in probabilistic logic programs. Ann. Math. Artif. Intell. (2012) (in press)
78. Straccia, U.: Fuzzy description logic programs. In: Proc. IPMU 2006, pp. 1818–1825 (2006)
79. Straccia, U.: Reasoning within fuzzy description logics. J. Artif. Intell. Res. 14, 137–166 (2001)
80. Straccia, U.: Towards a Fuzzy Description Logic for the Semantic Web (Preliminary Report). In: Gómez-Pérez, A., Euzenat, J. (eds.) ESWC 2005. LNCS, vol. 3532, pp. 167–181. Springer, Heidelberg (2005)
81. Straccia, U.: Uncertainty and description logic programs over lattices. In: Sanchez, E. (ed.) Fuzzy Logic and the Semantic Web, Capturing Intelligence, pp. 115–133. Elsevier (2006)
82. Swift, T.: Deduction in Ontologies via ASP. In: Lifschitz, V., Niemelä, I. (eds.) LPNMR 2004. LNCS (LNAI), vol. 2923, pp. 275–288. Springer, Heidelberg (2003)
83. Udrea, O., Yu, D., Hung, E., Subrahmanian, V.S.: Probabilistic Ontologies and Relational Databases. In: Meersman, R. (ed.) CoopIS/DOA/ODBASE 2005. LNCS, vol. 3760, pp. 1–17. Springer, Heidelberg (2005)
84. van Emden, M.: Quantitative deduction and its fixpoint theory. J. Log. Program. 3(1), 37–53 (1986)
85. van Keulen, M., de Keijzer, A., Alink, W.: A probabilistic XML approach to data integration. In: Proc. ICDE 2005, pp. 459–470. IEEE Computer Society (2005)
86. W3C. OWL Web Ontology Language Overview, 2004. W3C Recommendation (February10, 2004), `http://www.w3.org/TR/owl-features/`
87. W3C. OWL 2 Web Ontology Language Document Overview, 2009. W3C Recommendation (October 27, 2009), `http://www.w3.org/TR/owl2-overview/`
88. Yue, A., Liu, W., Hunter, A.: Measuring the Ignorance and Degree of Satisfaction for Answering Queries in Imprecise Probabilistic Logic Programs. In: Greco, S., Lukasiewicz, T. (eds.) SUM 2008. LNCS (LNAI), vol. 5291, pp. 386–400. Springer, Heidelberg (2008)

Author Index

MIX
Papier aus verantwortungsvollen Quellen
Paper from responsible sources
FSC® C105338

If you have any concerns about our products,
you can contact us on
ProductSafety@springernature.com

In case Publisher is established outside the EU,
the EU authorized representative is:
Springer Nature Customer Service Center GmbH
Europaplatz 3, 69115 Heidelberg, Germany

Printed by Libri Plureos GmbH
in Hamburg, Germany